14th Hungarian Conference on Materials Science 2023

IOP Conference Series: Materials Science and Engineering
Volume 1313

Balatonalmadi, Hungary
8-10 October 2023

ISBN: 979-8-3313-0660-1
ISSN: 1757-8981

Printed from e-media with permission by:

Curran Associates, Inc.
57 Morehouse Lane
Red Hook, NY 12571

Some format issues inherent in the e-media version may also appear in this print version.

Institute of Physics
Dirac House, Temple Back
Bristol BS1 6BE UK

Phone: 44 1 17 929 7481
Fax: 44 1 17 920 0979

techtracking@iop.org

Additional copies of this publication are available from:

Curran Associates, Inc.
57 Morehouse Lane
Red Hook, NY 12571 USA
Phone: 845-758-0400
Fax: 845-758-2633
Email: curran@proceedings.com
Web: www.proceedings.com

TABLE OF CONTENTS

Preface

Peer Review Statement

TIG Welding of In-Situ Produced Syntactic Metal Foam-Filled Tubes 1
A Kemény, G Pados, K Májlinger, I N Orbulov

Finite Element Modelling of Continuous Fiber–reinforced Composites Produced by Automated Manufacturing 10
B Szederkényi, I Rácz, N K Kovács, T Czigány

Flow Curve, Necking and Finite Elements 18
L Tatár, P Bereczki

Tensile Test and Finite Element Simulation of Coarse-Grained Thin Aluminium Sheets 28
A Szlancsik, D Varga

Unidirectional Fibre Reinforcement of Syntactic Metal Foams 35
B Szovák, J E Maróti, A Kemény, I N Orbulov

The Effect of Forming Conditions and Treatment Process Applied to the Material on Springback 42
J E Dessie, Z Lukács

Impregnation Problems of Aluminum Castings 50
M Réger, R Horváth, Er Fábian, T Réti

Thermomechanical-Chemical Devulcanization of Ground Tire Rubber 58
Á Görbe, A Kohári, I Z Halász-Kutasi, T Bárány

Some Considerations on the Contradictions of the Theory of Frank-Read Sources 68
János Dobránszky, Balázs Vero

Physical Simulations of the Tool Degradation in Friction Stir Welding 76
Nour El Imane Djimaoui, Yoni Adonyi, Jeroen De Backer, Árpád Kovács, László Kuzsella, Valéria Mertinger

Characterization of Friction Between Metal-Polymer Material Pairs with Different Speeds and Surface Pressures 88
R. Kohlhéb, R Horváth, M. Réger

Redefining the Strain Aged State of Unalloyed Mild Steels 94
B Kondas, V Mertinger

The Challenges of 3D Metal Printing and the Machinability Characteristics of Additively Produced Materials: A Review of Hybrid Machining 103
B Z Balazs, A Jacso, V Tancsa, M Takacs

Manufacturing of Composites with Designed Failure 111
G Zs Marton, Á Fendrik, G Szebényi

The Effect of Building Strategies on Tensile Strength of Additively Manufactured Parts by Laser Metal Deposition 120
G Juhász, F Hareancz, G Buza, R E Fábián, Á Vida

Investigation of Co-Cr Alloy Layer Prepared by Laser Metal Deposition 126
F Hareancz, G Juhász, G Buza, R E Fábián, Á Vida

Characterization of the Hydrostatic Core and Plastic Zone that Form During Brinell and Vickers Hardness Tests 131
F. Oláh, M. Réger, R Horváth

Improving the Bonding Technology of Aluminium Sheet Used in the Construction Industry by Gas Flame Surface Treatment 137
F Tajti, A Capár

Adhesive Bonding of Carbon Fiber Reinforced Polymer Composite 144
P I Kovács

Author Index

IOP Conf. Series: Materials Science and Engineering **1313** (2024) 011001 doi:10.1088/1757-899X/1313/1/011001

14th Hungarian Conference on Materials Science

Edited by

Z. Weltsch
Széchenyi István University, Hungary

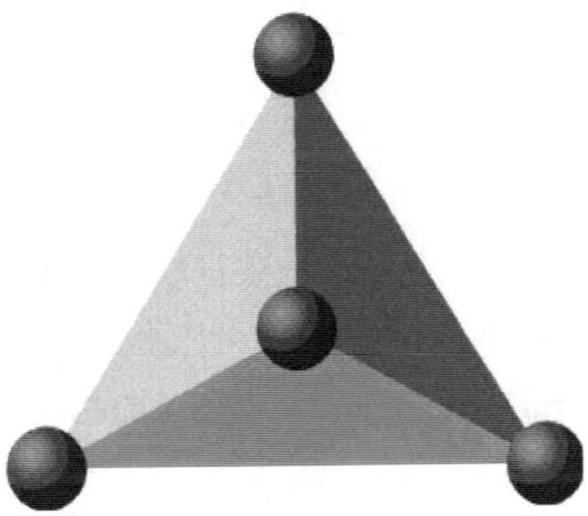

IOP Conf. Series: Materials Science and Engineering **1313** (2024) 011001		doi:10.1088/1757-899X/1313/1/011001

Scientific Committee

András Roósz – chairman

Tibor László Alpár
István Bársony
Dezső Beke
Károly Belina
Enikő Bitay
Zoltán Gácsi
Jenő Gubicza
Ákos Horváth
Richárd Horváth
Csaba Janáky
György Kaptay
Sándor Kéki
István László
János Lukács
Éva Majorosné Lublóy
Tamás Mankovits
Norbert Miskolczi
András Tompos
Péter Trampus
Ibolya Zsoldos

Organising Committee

Péter J. Szabó – chairman

Dénes Zsámbók honorary chairman

Tamás Bárány
Mihály Réger
Valéria Mertinger
György A. Gémes
Balázs Verő
Zoltán Weltsch

Conference Secretariat

Ildikó Kónya – secretary

e-mail: oatk@oatk.hu

Diamond Sponsors of the Conference

Struers

SZÉCHENYI
EGYETEMI CSOPORT
TAGJA

zone

Sponsor of the Conference

ZEISS

Hungarian Academy of Science

IOP Conf. Series: Materials Science and Engineering **1313** (2024) 011001 doi:10.1088/1757-899X/1313/1/011001

Preface

The series "Hungarian Conference and Exhibition on Materials Science, Testing and Informatics" (later simply "Hungarian Conference on Materials Science") was intended to provide a forum where Hungarian and foreign scientists and research groups interested in metals and alloys, silicates, polymers and composites get an opportunity to exchange and publish ideas and establish new and integrated partnerships. The 14th Hungarian Conference on Materials Science was held at the Balaton lakeshore in Balatonalmádi, October 8-10, 2023.

The winner of the photo competition of the 14th Hungarian Conference on Materials Science was Kovács Árpád Győző, University of Miskolc

The title of the picture: "Flower, my sweet flower".
The photo shows a fragment of a glass-fiber-reinforced polymer component (GF30), a degraded polymer matrix. The examination technique: ZEISS EVO MA 10 scanning electron microscope (HV mode, 20 kV, gold-coated surface, BSE image)

The organisers present this volume as a peer-reviewed selection of papers submitted to publication by participants

The organisers express their gratitude to all sponsors. Thanks are due to organising societies, as the Hungarian Materials Science Society and to other Hungarian scientific and engineering societies for their support.

Budapest, March, 2024.

Peter J. SZABÓ
Chairman of Organising Committee

IOP Conf. Series: Materials Science and Engineering **1313** (2024) 011002
doi:10.1088/1757-899X/1313/1/011002

Peer Review Statement

All papers published in this volume have been reviewed through processes administered by the Editors. Reviews were conducted by expert referees to the professional and scientific standards expected of a proceedings journal published by IOP Publishing.

- **Type of peer review:** Single Anonymous
- **Conference submission management system:** Morressier
- **Number of submissions received:** 22
- **Number of submissions sent for review:** 21
- **Number of submissions accepted:** 20
- **Acceptance Rate (Submissions Accepted / Submissions Received × 100):** 90.9
- **Average number of reviews per paper:** 2.2
- **Total number of reviewers involved:** 16
- **Contact person for queries:**
  Name: Zoltan Weltsch
  Email: weltsch.zoltan@sze.hu
  Affiliation: Szechenyi Istvan Egyetem

IOP Conf. Series: Materials Science and Engineering **1313** (2024) 012001 doi:10.1088/1757-899X/1313/1/012001

TIG welding of *in-situ* produced syntactic metal foam-filled tubes

A Kemény[1,2,*], G Pados[1,2], K Májlinger[1] and I N Orbulov[1,2]

[1] Department of Materials Science and Engineering, Faculty of Mechanical Engineering, Budapest University of Technology and Economics, Műegyetem rkp. 3., H-1111 Budapest, Hungary
[2] MTA-BME Lendület "Momentum" High-performance Composite Metal Foams Research Group, Műegyetem rkp. 3., H-1111 Budapest, Hungary

* E-mail: kemeny.alexandra@gpk.bme.hu

Abstract. Aluminium matrix double composites were produced via an *in-situ* pressure infiltration method. AlSi12 alloy matrix with densely packed lightweight expanded clay aggregate particle-filled foam was produced *in-situ* within AlMgSi0.5 alloy tubes. The attainable size of such double composites is limited; therefore, joining them is required to create larger parts. For the joining of these double composites, the applicability of tungsten inert gas (TIG) welding was tested. TIG welding was found to be applicable to join such composites. Metallurgically sound welded joints were produced. Nevertheless, the TIG welding of metal foam-filled tubes is very challenging and needs experienced welding personnel, and the weld bead was a bulk material, not a foam anymore.

1. Introduction

Metal foams are innovative materials with unique and beneficial properties, such as low density, high strength, good thermal and chemical stability and excellent energy absorption capabilities. Foam materials consist of a solid base material and gas phase voids (cells) that form the pores. The properties of the metal foams are generally influenced by (i) the material properties of the solid phase, (ii) the relative density of the foam (the ratio of the (bulk) density of the foam and the density of the matrix material), (iii) the spatial arrangement of the solid phase: the structure of the metal foam, and (iv) the properties of the second solid phase (filler) if applicable [1].

These materials are characterised by a cellular structure, which could include open [2] or closed cells [3], depending on the penetrability of the cells. Metal matrix syntactic foams (MMSFs) [4–6] are a subgroup of closed-cell foams consisting of a metal matrix and a hollow or porous filler material (ceramic of metal), where the size and shape of the latter are defined.

A bare foam is usually not the best solution to a particular engineering issue. Instead of basic foam panels, stiffness optimisation demands sandwich panels with dense face sheets or incorporating them into dense metallic sections such as tubes [7].

1.1. Metal foam-filled tubes

The fabrication of foam-filled tubes (FFTs) can be categorised as an *ex-situ* [8–11] or an *in-situ* [12–18] method, based on that the foam has been placed into the tube after the foam production or the foam has been produced into the tube in one step. However, in both cases, the strength results are significantly increased compared to the properties of the foam alone. With *in-situ* fabrication, the manufacturing process ensures seamless foam integration with the tube, optimising structural integrity.

On the other hand, the *ex-situ* method involves the insertion of a pre-fabricated metal foam into existing tube structures. This approach may offer more versatility in material choices and give a wide range of opportunities to improve the properties of the foam. However, it must be mentioned that the *ex-situ* method could be challenging to implement properly, considering the structural integrity.

The selection of *in-situ* or *ex-situ* methods depends on the specific application requirements, production efficiency, and desired material characteristics. FFTs have applications in fields such as automotive engineering and rail-based vehicles, where their unique combination of lightweight design and energy absorption capabilities is highly advantageous [7]. It is crucial to find a proper way to optimise welding and other joint technologies to increase the applicability of FFTs in the industrial area.

1.2. Joining of aluminium matrix composites

Due to the nature of the pressure infiltration process, the achievable size for the FFTs is very limited; therefore, using some kind of joining process to produce larger parts is inevitable. The highest joint strengths can be achieved via soldering, brazing, and welding. Aluminium matrix composite (AMC) soldering was already done on SiC particle-reinforced AMCs [19,20] and on closed-cell foams [21], but not on MMSFs. Brazing of aluminium alloys is also a pretty common industrial process. SiC particle-reinforced AMCs were joined with different brazing techniques [22–24], and aluminium foams were also successfully brazed [25], but to our knowledge, no MMSFs have been joined by welding. Undoubtedly, the highest achievable joint strength for metals is offered by welding. There are a lot of welding processes which are used to weld AMCs.

Solid-state welding, as well as fusion welding techniques, were applied to join AMCs. Linear friction welding was used to join SiC particle-reinforced AMCs [26]. Friction stir welding (FSW) is an advantageous technique for hard-to-weld aluminium alloys such as Li-alloyed ones. The process was invented by The Welding Institute (TWI) of Cambridge, England, in 1991. Since then, the welding process found its way to every industry branch. Many AMCSs with different reinforcing particle types (SiC, Zrb_2, AlN, B_4C, Al_2O_3 and different nanoparticles) were welded successfully with FSW [27–29]. FSW was even used to produce the precursor for aluminium alloy foam core sandwich panels [30–33]. One major disadvantage compared to the FSW of pure aluminium is that in the case of AMCs, the tool wear is very rapid and severe, increasing with the reinforcing particle amount [34]. FSW was also used for welding aluminium foams [35]. Another solid-state welding, the impact welding method was used to join SiC particle-reinforced AMCs [36].

Fusion welding techniques are also used for the joining of AMCs. Different particles (ZrB_2, TiB_2, SiC) reinforced AMCs were successfully welded via laser beam welding [37–39] and also aluminium foams [35]. Cambronero et al. [40] even used concentrated solar energy to weld aluminium foams with precursor bars of AlSi (10–12% Si) alloy with 0.8% TiH_2 as the foaming agent was used as filler material. So, the weld bead itself actually became a foam. Metal inert gas (MIG) welding was also applied to join SiC, Al_2O_3, and particle-reinforced AMCs [35,41] and also for joining aluminium foam plates [35,42]. Of course, due to the solid Al alloy filler wire, the weld metal itself wasn't a composite anymore. Similarly to MIG welding, tungsten inert gas (TIG) welding was also successfully applied in the joining of particle-reinforced AMCs (SiC, Al_2O_3, B_4C, even with nanoparticles and rice husk ash-filled ones) [41,43–46]. TIG welding was also used to join aluminium foam plates [35]. As can be seen, a lot of welding techniques are available to join AMCs, but to our best knowledge, there is no research published where lightweight expanded clay aggregate particle (LECAP)-filled MMSF was welded, not to mention LECAP-filled FFTs.

Therefore our aim was to investigate the feasibility of TIG welding to join LECAP filled FFTs.

2. Materials and methods

2.1. Production of the syntactic metal foam-filled tubes

In this research, *in-situ* FFTs were produced with an MMSF core via the low-pressure infiltration technique, similarly as in [47]. The tube had a wall thickness of 2 mm and an outer diameter of Ø32 mm and was made of AlMgSi0.5 alloy. The matrix material of the metal foam core was a near-eutectic casting alloy (AlSi12), and the filler material was ⌀2.5–3.0 mm sized LECAPs, obtained from Liapor

GmbH & Co. KG (Hallerndorf, Germany). The chemical composition of the particles has been given in weight percentage by the supplier as: ~60 % SiO_2, ~17 % Al_2O_3, ~14 % Fe_2O_3 and ~9 % other oxides. Table 1 shows the chemical composition of the applied aluminium alloys, measured by a WAS PMI MasterSort optical emission spectroscope. The infiltration parameters have been detailed in a previous research paper by the authors [18].

Table 1. The standardised chemical composition of the used materials in wt. %.

	Al	Si	Mg	Fe	Mn	Ti	Other
AlSi12	86.6 ± 2.0	12.5 ± 1.7	<0.001	0.50 ± 0.06	0.15 ± 0.02	0.14 ± 0.01	0.07 ± 0.10
AlMgSi0.5	99.2 ± 0.0	0.42 ± 0.06	0.19 ± 0.05	0.14 ± 0.02	<0.001	0.01 ± 0.01	0.04 ± 0.03

2.2. Welding of the FFTs

Prior to the welding, the FFTs were chamfered to get access with the welding torch. The machined geometry and the prepared specimens can be seen in Fig. 1. Furthermore, it can be seen in Fig. 1b and c that, however, the filler material is densely packed within the tube; on the prepared surfaces, its distribution is inhomogeneous due to its size compared to the machined surface.

Manual TIG welding was made with a Syrius® WIG 351 AC/DC PULSE welding machine. The shielding gas was pure Ar (99.996 %) with a 10 l/min flow rate. The tungsten electrode used was a Ø1.6 mm, grey (WC20 – 2 % CeO_2) electrode with a rounded tip. The welding current was 80–90 A, AC (50 Hz), and arc voltage was 15-16 V. Welding speed varied between 150–200 mm/min. The welding filler was a Ø2.4 mm AlSi5 (4043) aluminium alloy rod, with the nominal chemical composition of 4043 is 5 % Si, 0.8 % Fe, 0.3 % Cu, 0.05 % Mn, 0.05 % Mg, 0.1 % Zn, 0.2 % Ti and Al as balance. The root gap was set to ~5 mm. The welding was done in a flat position (PA) with prism support, but the welding direction was in the axial direction. First, a quasi-buttering layer was welded over the surface of the machined FFTs and after that, filler layers were welded in the axial direction. The FFT was rotated within the prism after each welded bead.

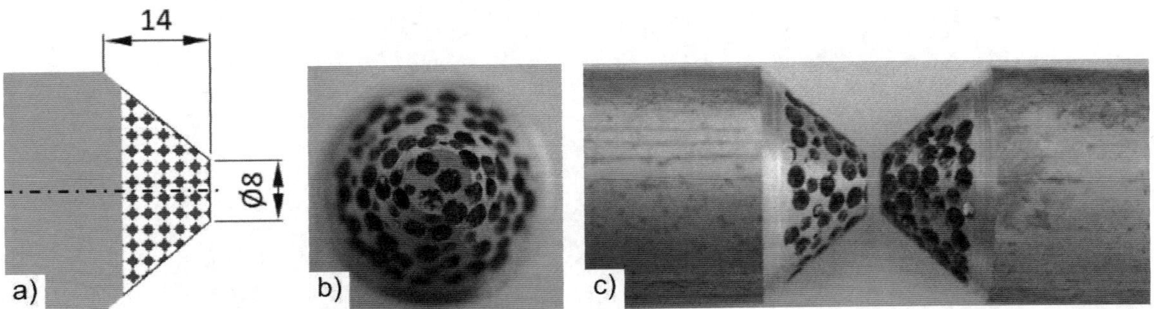

Figure 1. Schematic drawing of the specimen's edge preparation for welding (a) and images of the machined specimens in front view (b) and side view (c).

2.3. Investigations

Specimens were cut in half with a precision cutting machine at the length axis of the tubes and mounted in cold mounting epoxy resin. Standard metallographic preparation was done with continuous water rinsing till P4000 emery SiC sandpaper, followed by diamond suspension polishing with 3 μm and 1 μm grain size. Keller's reagent (95 ml H_2O, 2.5 ml HNO_3, 1.5 ml HCl and 1 ml HF) was used for the microscopy examinations. An Olympus SZX-6 type stereo microscope was used for the macrostructural images, and an Olympus PMG-3 type microscope was used for the microstructural images. Hardness measurements were made with a Buehler IndentaMet 1011 micro-Vickers hardness testing machine with a 200 g indentation load and 11 s indentation time on the cross-sectional specimens prepared for metallography.

3. Results and discussion

3.1. Welding of the FFTs

During the manual TIG welding, some problems arose that needed to be solved. The LECAP filler particles have significantly worse electrical conductivity (basically are dielectrics) than the aluminium alloys used; therefore, it caused: a) involuntary arc movement on the FFTs surface, and b) the LECAPs were significantly overheated compared to the matrix and filler material (based on their colour during welding ~850 – 1000°C can be estimated). *Note that the excess arc movement may be significantly reduced by using smaller-sized filler particles compared to the electric arc diameters.*

A different approach was used to solve the abovementioned problems than what is used in common tube welding or circular seam welding. The welding was done in a flat position, but the welding direction was in the axial direction. First, a quasi-buttering layer was welded over the surface of the machined FFTs with very short bead lengths (~ 10 mm). It was done because however small the solidification range of the filler and matrix material was, the overheated LECAP particles began to move within the melt pool. Due to their low density, they tried to move to the surface of the weld. After this quasi-buttering, the filler layers were also welded in an axial direction because of the unevenness of the first layers caused by the welding difficulties. *Note that the first quasi-buttering layer could have been machined to an even surface, and after that, circular welding could be made, but it would be very time-consuming and costly.*

The skilled welding professional compensated for the arc movement and the moving of the LECAPs effectively; therefore, a seemingly good joint was produced. The image of the welded joint surface can be seen in Fig. 2., with no visible surface defects. *Note that the mentioned difficulties will disable the use of automated TIG welding processes because the welder needs to react to the filler particle movement immediately. A prior aluminium coating on the surface could help with that.*

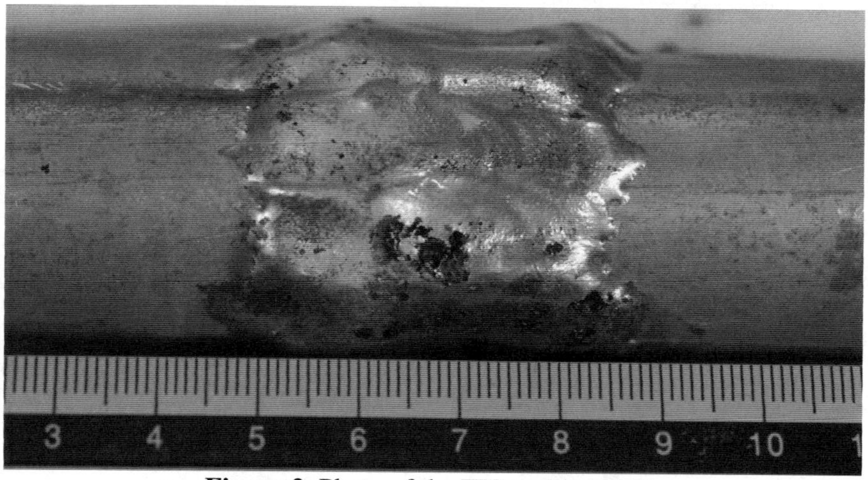

Figure 2. Photo of the TIG welded FFT.

3.2. Characterisation of the welded joint

In Fig. 3a, a macrostructural photo of the cross-section of the TIG welded FFT can be seen with the edge preparation indicated by red dashed lines. As it was observed during welding, it is also evident in this image that the spheres moved within the melt pool; one even almost reached the surface of the joint (Fig. 3b). On the upper part of the image, the detachment of the foam core from the tube wall can be observed (Fig. 3b-c) with the length of ~ 15 mm from the weld bead. This might be due to the difference in the thermal expansion of the foam and tube, but it can also be from prior defects that might have occurred during the FFTs' manufacturing. A small amount of gas inclusions were also observed (Fig. 3b-c) which is very common during aluminium welding. In the midsection in the root gap (Fig. 3c), some lack of fusion defects can be observed; that can be due to the movement of the spheres in the first layer of the weld bead. Within the void, two LECAP sphere parts can be seen.

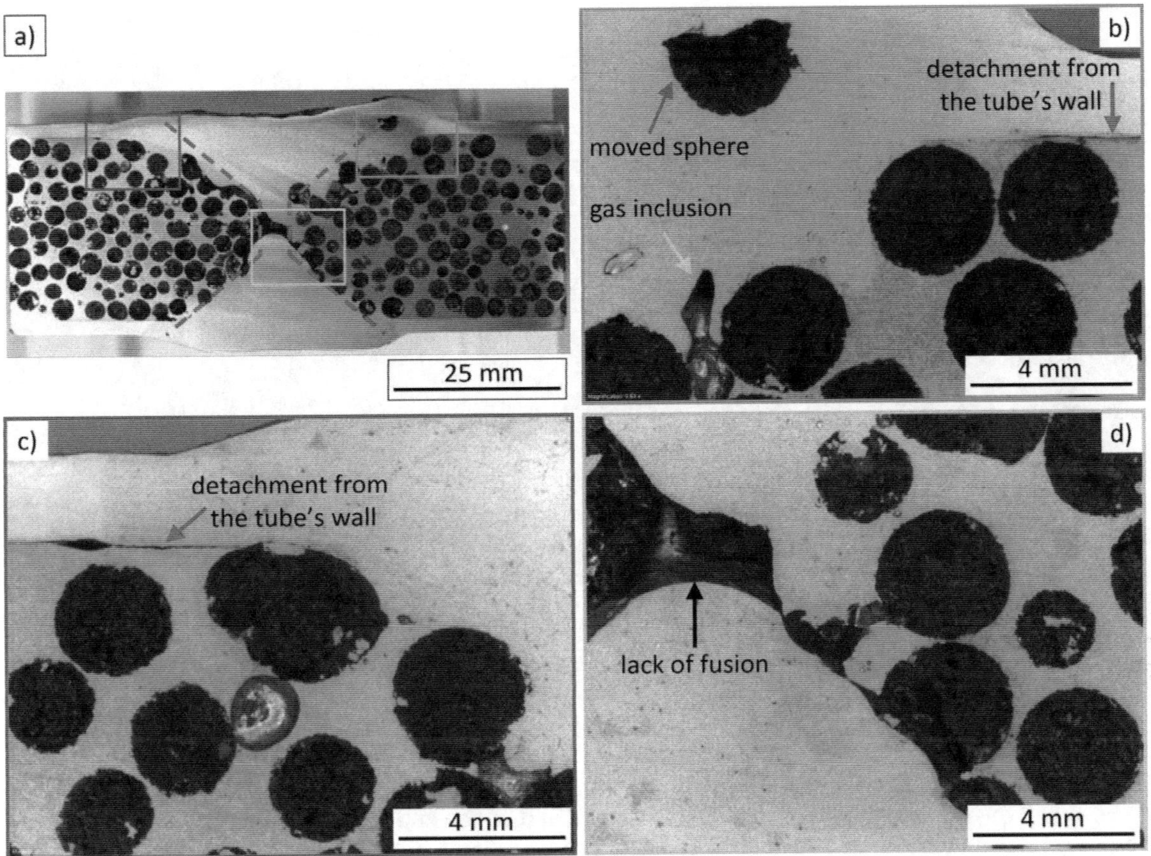

Figure 3. Macrostructural photos: a) of the cross-section of the TIG welded FFT where edge preparation for welding is indicated with red dashed line, and the areas for the enlarged images c)-d) are indicated in coloured rectangles.

Fig. 4. shows the micrographs of the different parts of the welded FFTs. The matrix (Fig. 4a) and the weld (Fig. 4b) represent different silicon distributions in the aluminium due to the different amounts of silicon in each alloy. The used magnification does not show much detail in the tube material (Fig. 4c), as it has a low amount of alloying elements that are present in fine precipitates. However, there is a significant mixed zone between the matrix and filler material with enriched silicon content on the weld material side due to the dilution of the two materials (Fig. 4d). In the case of the tube–weld border, there is a clear transition between the weld and the tube material (Fig. 4e).

Microhardness measurements were made along the mid-line of the weld bead to evaluate the homogeneity of the weld. Fig. 5 shows the hardness distribution in the mid-line of the weld zone, from the surface to the inside of the FFT. The hardness values of the AlSi5 weld had a surprisingly small scatter for a weld; the average is 49±4 HV0.2. The average hardness of the AlMgSi0.5 tube was 57±2 HV0.2, which can be increased with precipitation-hardening heat treatment. The AlSi12 matrix had an average hardness of 66±2 HV0.2, which is significantly higher than the achieved hardness of the weld; however, the mechanical effect of this difference should be investigated with bend tests and tensile tests, as the behaviour of the FFT is impacted by LECAPs, not only the matrix material. The average hardness of the diluted zone was 56±2 HV0.2. Another set of hardness measurements was made along the midline of the tube, from the edge of the "moved" sphere in Fig 3. b) (Fig. 6). Here, a softening in the heat affected zone (HAZ) can be observed, with a small ~ 1 mm HAZ width.

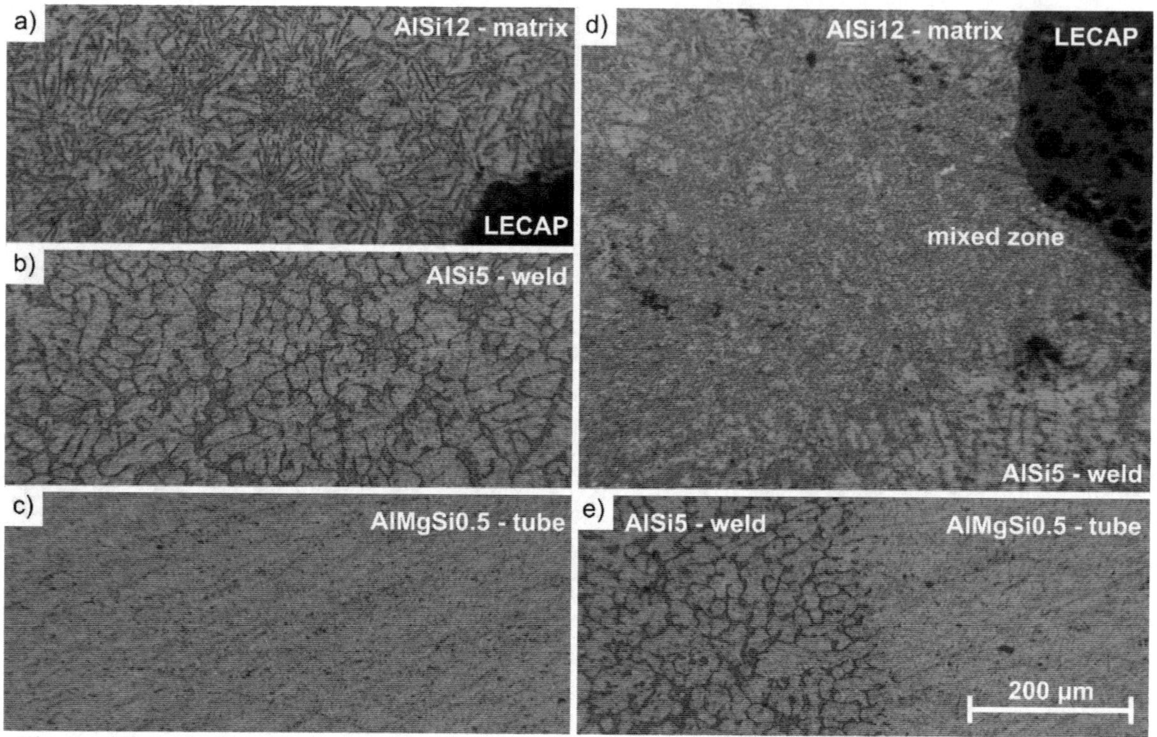

Figure 4. Microscopy images of the different parts of the joint: a) matrix material far from the weld, b) middle of the weld, c) tube material, d) enriched silicon in the mixed zone of the matrix and the weld and e) transition between the tube and the weld.

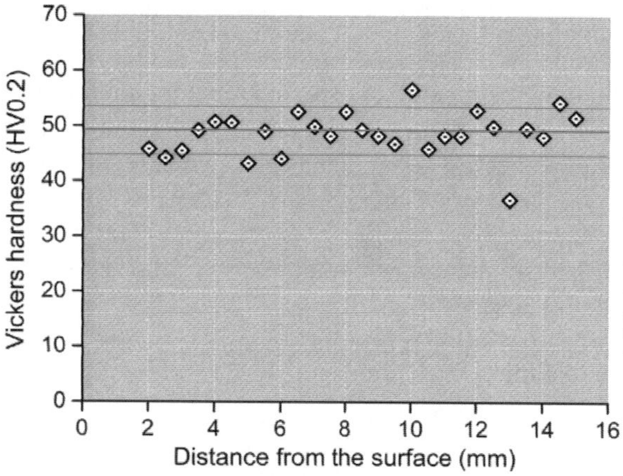

Figure 5. Hardness values in the midline of the weld bead, where the '0' distance corresponds to the outer tube wall. Mean hardness and scatter are indicated with red horizontal lines.

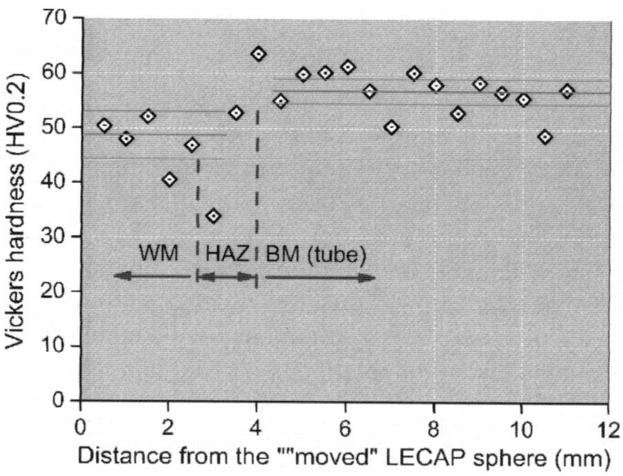

Figure 6. Hardness values in the midline of the tube wall, where the '0' distance corresponds to the edge of the "moved" sphere in Fig. 3b. Mean hardness and scatter for the weld metal and the tube are indicated with red horizontal lines.

IOP Conf. Series: Materials Science and Engineering **1313** (2024) 012001 doi:10.1088/1757-899X/1313/1/012001

4. Conclusions

The following conclusions were drawn from the experiment:

- Manual TIG welding is applicable to join aluminium matrix LECAP particle foam-filled tubes.
- Coarse filler particles can cause involuntary arc movement on the specimen.
- Filler particles were significantly overheated compared to the filler material, causing them to move around in the melt pool.
- Only very skilled welding personnel can compensate for the particle movement in the melt pool; therefore, commercially automated TIG welding techniques are not applicable to join these FFTs without any surface treatment, e.g. coating.
- A significantly diluted zone was found between the matrix and the weld material.
- The AlSi5 weld metal had a lower hardness than the AlSi12 matrix or the AlMgSi0.5 tube material, but as a bulk material, not a foam; therefore, the mechanical strength of the joint is more than satisfactory.

A further aim of the authors is to define a technology window for the welding possibilities of metallic FFTs, in the function of the properties of the tube (e.g. diameter, wall thickness and material), as well as the foam properties (e.g. cell structure, matrix and filler materials and relative density).

Acknowledgements

The research reported in this paper is part of project no. TKP-6-6/PALY-2021, which has been implemented with the support provided by the Ministry of Culture and Innovation of Hungary from the National Research, Development and Innovation Fund, financed under the TKP2021-NVA funding scheme. G. Pados was supported by the ÚNKP-23-2-I-BME-180 New National Excellence Program of the Ministry for Culture and Innovation from the source of the National Research, Development and Innovation Fund. This work was partially supported by the National Research, Development and Innovation Office (NKFIH), under grant agreement OTKA-FK_21 138505.

References

[1] Ashby M F, Evans A, Fleck N A, Gibson L J, Hutchinson J W and Wadley H N . 2002 Metal foams: a design guide *Mater. Des.* **23** 119
[2] Kapłon H, Blawert C, Chęcmanowski J and Naplocha K 2022 Development of open-porosity magnesium foam produced by investment casting *J. Magnes. Alloy.* **10** 1941–56
[3] Naeem M A, Gábora A and Mankovits T 2020 Influence of the Manufacturing Parameters on the Compressive Properties of Closed Cell Aluminum Foams *Period. Polytech. Mech. Eng.* **64** 172–8
[4] Kádár C, Kubelka P and Szlancsik A 2023 On the compressive properties of aluminum and magnesium syntactic foams: Experiment and simulation *Mater. Today Commun.* **35** 106060
[5] Mirzarahimi M K, Khodarahmi H, Ziashamami M and Hosseini R 2023 Numerical and Experimental Study of the Use of Mineral Pumice in the Core of the Sandwich Panel to Absorb the Shock Wave *Period. Polytech. Mech. Eng.* **67** 19–29
[6] Bálint A and Szlancsik A 2015 Mechanical Properties of Iron Hollow Sphere Reinforced Metal Matrix Syntactic Foams *Mater. Sci. Forum* **812** 3–8
[7] Banhart J 2007 Metal foams - From fundamental research to applications *Frontiers in the design of materials* (Taylor & Francis) pp 279–89
[8] Movahedi N and Linul E 2017 Quasi-static compressive behavior of the ex-situ aluminum-alloy foam-filled tubes under elevated temperature conditions *Mater. Lett.* **206** 182–4
[9] Linul E, Movahedi N and Marsavina L 2017 The temperature effect on the axial quasi-static compressive behavior of ex-situ aluminum foam-filled tubes *Compos. Struct.* **180** 709–22
[10] Su M, Wang H and Hao H 2019 Axial and radial compressive properties of alumina-aluminum matrix syntactic foam filled thin-walled tubes *Compos. Struct.* **226** 111197

IOP Conf. Series: Materials Science and Engineering **1313** (2024) 012001 doi:10.1088/1757-899X/1313/1/012001

[11] Garai F, Béres G and Weltsch Z 2020 Development of tubes filled with aluminium foams for lightweight vehicle manufacturing *Mater. Sci. Eng. A* **790** 139743

[12] Duarte I, Krstulović-Opara L and Vesenjak M 2015 Characterisation of aluminium alloy tubes filled with aluminium alloy integral-skin foam under axial compressive loads *Compos. Struct.* **121** 154–62

[13] Duarte I, Vesenjak M, Krstulović-Opara L, Anžel I and Ferreira J M F 2015 Manufacturing and bending behaviour of in situ foam-filled aluminium alloy tubes *Mater. Des.* **66** 532–44

[14] Duarte I, Krstulović-Opara L and Vesenjak M 2018 Axial crush behaviour of the aluminium alloy in-situ foam filled tubes with very low wall thickness *Compos. Struct.* **192** 184–92

[15] Taherishargh M, Vesenjak M, Belova I V, Krstulović-Opara L, Murch G E and Fiedler T 2016 In situ manufacturing and mechanical properties of syntactic foam filled tubes *Mater. Des.* **99** 356–68

[16] Fiedler T, Moore R and Movahedi N 2021 Manufacturing and characterization of tube-filled ZA27 metal foam heat exchangers *Metals (Basel).* **11**

[17] Chilla V, Mondal D P, Ram G D J and Mukherjee M 2022 Processing of in-situ aluminium foam-filled stainless steel tube with foam-tube bonding for enhanced crashworthiness *J. Manuf. Process.* **82** 488–500

[18] Kemény A, Leveles B, Kincses D B and Károly D 2022 Manufacturing and Investigation of In Situ and Ex Situ Produced Aluminum Matrix Foam-Filled Tubes *Adv. Eng. Mater.* **24** 2100365

[19] Lu J, Mu Y, Luo X and Niu J 2012 A new method for soldering particle-reinforced aluminum metal matrix composites *Mater. Sci. Eng. B* **177** 1759–63

[20] Nowacki J and Moraniec K 2015 Evaluation of Methods of Soldering AlSi and AlSi-SiC Particle Composite Al Foams *J. Mater. Eng. Perform.* **24** 426–33

[21] Huang Y, Gong J, Lv S, Leng J and Li Y 2012 Fluxless soldering with surface abrasion for joining metal foams *Mater. Sci. Eng. A* **552** 283–7

[22] Zhang X P, Quan G F and Wei W 1999 Preliminary investigation on joining performance of SiCp-reinforced aluminium metal matrix composite (Al/SiCp–MMC) by vacuum brazing *Compos. Part A Appl. Sci. Manuf.* **30** 823–7

[23] Niu J, Luo X, Tian H and Brnic J 2012 Vacuum brazing of aluminium metal matrix composite (55vol.% SiCp/A356) using aluminium-based filler alloy *Mater. Sci. Eng. B* **177** 1707–11

[24] Ferraris M, Gili F, Lizarralde X, Igartua A, Mendoza G, Blugan G, Gorjan L and Casalegno V 2022 SiC particle reinforced Al matrix composites brazed on aluminum body for lightweight wear resistant brakes *Ceram. Int.* **48** 10941–51

[25] Nannan C, Yi F, Jie C, Bin L, Fanyan C and Jingsong Z 2013 Vacuum Brazing Processes of Aluminum Foam *Rare Met. Mater. Eng.* **42** 1118–22

[26] Rotundo F, Marconi A, Morri A and Ceschini A 2013 Dissimilar linear friction welding between a SiC particle reinforced aluminum composite and a monolithic aluminum alloy: Microstructural, tensile and fatigue properties *Mater. Sci. Eng. A* **559** 852–60

[27] Salih O S, Ou H, Sun W and McCartney D G 2015 A review of friction stir welding of aluminium matrix composites *Mater. Des.* **86** 61–71

[28] Bahrami M, Dehghani K and Besharati Givi M K 2014 A novel approach to develop aluminum matrix nano-composite employing friction stir welding technique *Mater. Des.* **53** 217–25

[29] Kurt H I, Oduncuoglu M and Asmatulu R 2016 Wear Behavior of Aluminum Matrix Hybrid Composites Fabricated through Friction Stir Welding Process *J. Iron Steel Res. Int.* **23** 1119–26

[30] Hangai Y, Kamada H, Utsunomiya T, Kitahara S, Kuwazuru O and Yoshikawa N 2014 Aluminum alloy foam core sandwich panels fabricated from die casting aluminum alloy by friction stir welding route *J. Mater. Process. Technol.* **214** 1928–34

[31] Hangai Y, Takada K, Endoh R, Koyama S and Utsunomiya T 2017 Foaming behavior of aluminum foam precursor induced by friction heat generated by rotating tool used for spot friction stir welding *J. Manuf. Process.* **25** 426–31

[32] Hangai Y, Takada K, Fujii H, Aoki Y and Utsunomiya T 2019 Foaming behavior of blowing-

and stabilization-agent-free aluminum foam precursor during spot friction stir welding *J. Mater. Process. Technol.* **265** 185–90

[33] Hangai Y, Takada K, Endo R, Fujii H, Aoki Y and Utsunomiya T 2018 Foaming of aluminum foam precursor during friction stir welding *J. Mater. Process. Technol.* **259** 109–15

[34] Bist A, Saini J S and Sharma B 2016 A review of tool wear prediction during friction stir welding of aluminium matrix composite *Trans. Nonferrous Met. Soc. China* **26** 2003–18

[35] Feng M, Ren L, Wang Z, Meng D, Wang S and Wang P 2023 Recent research progress of foam metals welding: a review *Int. J. Adv. Manuf. Technol.* **127** 3135–56

[36] Mao Y, Li J, Vivek A and Daehn G S 2021 High strength impact welding of 7075 Al to a SiC-reinforced aluminum metal matrix composite *Mater. Lett.* **303** 130549

[37] David Raja Selvam J, Dinaharan I and Mashinini P M 2021 Microstructure and mechanical characterization of Nd:YAG laser beam welded AA6061/10 wt% ZrB2 aluminum matrix composites *Opt. Laser Technol.* **140** 107084

[38] Meng C, Cui H, Lu F and Tang X 2013 Evolution behavior of TiB2 particles during laser welding on aluminum metal matrix composites reinforced with particles *Trans. Nonferrous Met. Soc. China* **23** 1543–8

[39] Wang Z, Cao H, Li H, Wang D, Xia H, Butt H A, Li M and Liu D 2023 Effects of laser welding parameters on the porosity and acicular phase in SiCp/6092 aluminum matrix composite welded joints *J. Mater. Res. Technol.* **23** 5127–41

[40] Cambronero L E G, Cañadas I, Ruiz-Román J M, Cisneros M and Corpas Iglesias F A 2014 Weld structure of joined aluminium foams with concentrated solar energy *J. Mater. Process. Technol.* **214** 2637–43

[41] Vigneshwar M, Selvamani S T, Tarun K and Palanikumar K 2019 A novel approach for joining armor grade AA7075 metal matrix nano composites using various welding processes *Mater. Today Proc.* **16** 1175–81

[42] Shih J S, Tzeng Y F and Yang J Bin 2011 Principal component analysis for multiple quality characteristics optimization of metal inert gas welding aluminum foam plate *Mater. Des.* **32** 1253–61

[43] Jayashree P K, Basu R and Sharma S S 2021 An electron backscattered diffraction (EBSD) approach to study the role of microstructure on the mechanical behavior of welded joints in aluminum metal matrix composites *Mater. Today Proc.* **38** 490–3

[44] Kalaiselvan K, Dinaharan I and Murugan N 2014 Characterization of friction stir welded boron carbide particulate reinforced AA6061 aluminum alloy stir cast composite *Mater. Des.* **55** 176–82

[45] Saini S, Singh S, Singh K and Singh A 2020 Some studies into weldability of rice husk ash aluminium matrix composites using TIG welding *Mater. Today Proc.* **24** 298–307

[46] Senniangiri N, Velliyangiri B, PavinRaj S, Sakthivel S, Sathiyanarayanan K and Vishnu Vardhan B S 2023 Investigation on the impact of B4C content in AA5052 based metal matrix composites on the mechanical properties of welded joints *Mater. Today Proc.*

[47] Pados G, Kemény A, Szovák B, Károly D and Orbulov I N 2023 Fabrication of In-situ Syntactic Aluminium Foam-Filled Steel Tubes *Acta Mater. Transylvanica* **6** 38–41

IOP Conf. Series: Materials Science and Engineering **1313** (2024) 012002
doi:10.1088/1757-899X/1313/1/012002

Finite element modelling of continuous fiber–reinforced composites produced by automated manufacturing

B Szederkényi[1], **I Rácz**[1], **N K Kovács**[1,2] **and T Czigány**[1,3*]
[1] Department of Polymer Engineering, Faculty of Mechanical Engineering, Budapest University of Technology and Economics, Muegyetem rkp. 3., H-1111 Budapest, Hungary
[2] MTA-BME Lendület Lightweight Polymer Composites Research Group, Muegyetem rkp. 3., H-1111 Budapest, Hungary
[3] HUN-REN-BME Research Group for Composite Science and Technology, Muegyetem rkp. 3., H-1111 Budapest, Hungary
* corresponding author: czigany@eik.bme.hu

Abstract. With the advent of automated, continuous filament placement technologies, the possibilities for reinforcement placement in composite manufacturing have been further expanded, as the reinforcement path can be continuously varied within the layer. This allows far more efficient structures to be created, resulting in a further reduction in the weight of the structure compared to conventional composites. At the same time, this manufacturing freedom also poses a major challenge for the structural design of composite components, as simulation methods based on meso-level homogenization are mostly not usable due to the lack of periodicity of the reinforcement structure. In our paper, we have established a structural modelling method using finite element analysis (FEA) for fused filament fabrication-based additive manufacturing of continuous fiber–reinforced composites. For modelling the reinforcement structure properly, we used the base-coordinate sweep (BCS) modelling method and for capturing progressive failure mechanisms we used Ansys LS-Dyna with the MAT54 material card. Based on the simulations and tests performed, we demonstrated that the method we present is suitable for the engineering modelling of continuous fiber–reinforced 3D printed composites.

1. Introduction

As the world changes and evolves, engineers are increasingly driven to use and develop more resource- and energy-efficient structural elements. Achieving this goal requires the optimization and proper design of lightweight structural materials. Composites are the structural materials with the highest specific strength currently known, and to exploit their advantageous properties, it is necessary to orient the reinforcing fibers in the direction of the load. In this context, additive manufacturing has opened up new possibilities for designers. Composite 3D printing offers geometric freedom in fiber placement that would have been previously unimaginable or very difficult to achieve with conventional manufacturing technologies. However, exploiting this novel method also requires novel tools and methods to optimize complex geometries and fiber reinforcement structures. Topology Optimization (TO) satisfies this requirement, allowing engineers to find the material distribution that provides the best performance with the least material usage [1].

This paper aims to develop a procedure for structural analysis based on finite element methods (FEM) for these novel composite manufacturing technologies, including the modelling of progressive failure mechanisms.

IOP Conf. Series: Materials Science and Engineering **1313** (2024) 012002 doi:10.1088/1757-899X/1313/1/012002

2. Challenges in the structural analysis of composites produced by automatic fiber placement

The structural analysis of continuous fiber-reinforced 3D printed composites is a challenge in design and manufacturing due to several factors. Exploiting the intricate anisotropic properties arising from the variable orientation of the reinforcement fibers plays a key role in optimizing the components, but there is currently no widespread method for automatically incorporating these into FE models. Beyond software limitations, the situation is complicated by manufacturing specificities and current technology limitations. In the case of fused filament fabrication (FFF) based printing methods the bulk lay-up is welded in a way that it does not fill the available space due to the extruder profile [2,3]. Porosity, and interlayer welding problems during printing have a non-negligible influence on the mechanical properties of these composite parts. As porosity can exceed 10%, the failure of the part is highly dependent on the generation and propagation of micro- and meso-structural defects in porous parts under load [4]. Further difficulties are caused by the constantly evolving and therefore often changing material composition, which also requires the continuous updating of existing models to adequately model failure phenomena [5]. A key part of our work was searching for and analyzing existing material model sources and performing our validation tests. The simulation and numerical modelling of 3D printed composites is a complex, multi-level problem that requires a different approach from traditional methods.

3. Modelling of composites produced by automated fiber placement

The meso-level computational methods developed for conventional composites based on layer homogenization cannot be used for complex reinforcement structures, since the reinforcement layer cannot be described with a finite number of discrete vectors for principal directions. Instead, the mechanical behavior of composites can be more accurately described with an orientation-dependent ($\underline{\underline{M}}(\theta)$) elastic matrix, where θ represents the local principal direction of reinforcement. For conventional systems, such as unidirectionally reinforced or multi-directionally reinforced fabrics with repetitive patterns, the elasticity matrix values remain constant across the whole layer. However, with automated fiber placement, these θ values could vary locally within the layer, making the structure more adaptable to specific loading conditions. This variability enhances the specific performance of the composite structure, but also eliminates traditional homogenization methods at the layer level, necessitating new modelling approaches (Figure 1: (a)). In our paper, we address this problem and present a possible solution using the base-coordinate sweep (BCS) method. The BCS procedure can consider the change of fiber orientation within the layer by treating the reinforcement bundles as the lowest level of the structure. By fitting a coordinate system running along the centerline of the fiber bundles (Figure 1: (b)), the BCS method can provide a more accurate prediction of the processes occurring in the component.

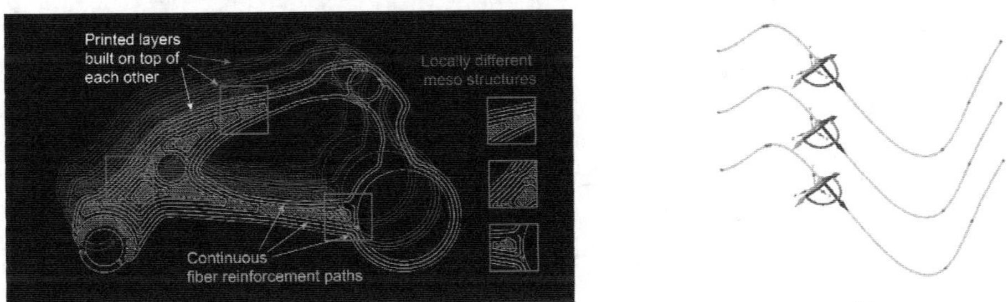

(a) (b)

Figure 1. Typical structure of composites produced with automated methods on a Formula Student car suspension part, where the meso-structure does not allow homogenization at the layer level (a) and a schematic diagram showing the base-coordinate sweep along the reinforcement curves (b)

To validate our approach, we chose a geometry from the literature [6] known to be generated by TO, thus allowing the use of complex reinforcement trajectories. Our previous paper [7] had already addressed this geometry, where we developed optimal reinforcement trajectories for the component and introduced a fiber segmentation method, which allowed us to model the stiffness of the component correctly but did not provide any information on the post-failure process. In our current paper, we continue to work with this reinforcement structure (Figure 2).

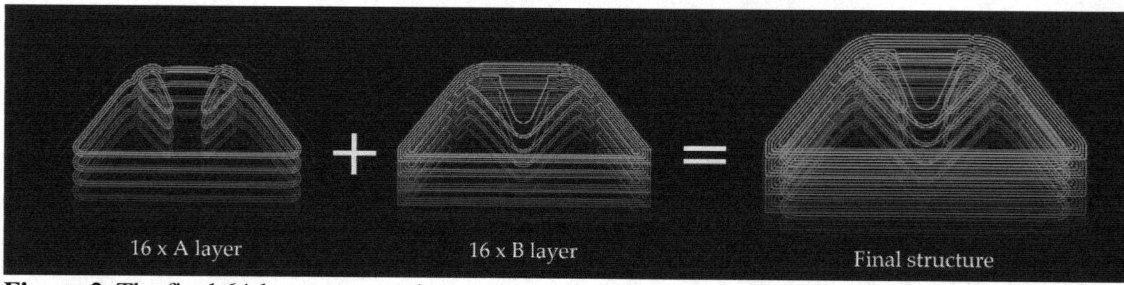

Figure 2. The final 64-layer test specimen with a reinforcement structure consisting of 2x16 alternating separate reinforcement layers and "blank" layers printed in between, containing only matrix material.

4. Manufacturing technology, materials used, material model, and the measurement environment

The composite part was 3D printed with a Markforged Mark Two FFF 3D printer, set to a layer thickness of 0.125 mm. The matrix material of the composite part was Markforged Onyx base material (PA6 filled with milled carbon fiber), the continuous filament bundle was 1K roving (1000 elementary fibers) pre-impregnated with PA6, with a fiber content of 28% and a filament diameter of 0.375 mm, taken from the literature [8,9].

For our simulations, the initial material properties of the components were taken from the literature [8,10–12], supported by our tests in the case of continuous carbon filament. However, due to the limitations of the Markforged system, which does not allow the components to be manufactured completely separated, the results obtained had to be corrected afterward to get the properties of individual components. These corrections were performed with the use of the rule of mixtures, according to Eq. (1).

$$E_f = \frac{E_{resultant} - (v_m E_m + v_v E_v)}{v_f} = 70{,}65\ [GPa] \qquad (1)$$

where ***E*** is the tensile modulus of elasticity [GPa], ***V*** is the volume fraction [-], ***m*** is an index of the matrix material, ***f*** is an index of the fiber-reinforced filament, and ***v*** is an index of voids in the part, also known as the porosity of the structure. The volume fractions of the components were calculated from the cross-sectional ratios measured in our specimens and the known diameter of the fiber-reinforced filament, with simple geometric relationships.

The modulus-type and Poisson's ratio values in a non-principal direction, which cannot be determined indirectly from tensile and three-point bending tests (Figure 3. a and b), were determined with the use of the micromechanical representative volume element (RVE) modelling method, where the stiffness and principal direction strength of the filament was determined from the fiber content and Poisson's ratio of the components. Then, the shear modulus and Poisson's ratio of the continuous fiber-reinforced orthotropic UD filament (Figure 3. c) were determined by tuning the output to the measured values. Strength properties perpendicular to the principal direction, and shear strength were estimated with the values of the matrix material found in the literature [13]. The results are summarized in Table 1 (a).

The bending test was carried out on a Zwick Z250 universal testing machine, at room temperature, at a test speed of 2mm/min, on 3 specimens. We designed a fixture for the test, specifically for this non-standard specimen geometry (Figure 3 a, b). These tests were complemented by Digital Image Correlation (DIC) to track the local deformation fields of the specimen. This provided us with more accurate visual information on the progress of failure, which we later compared with the simulation.

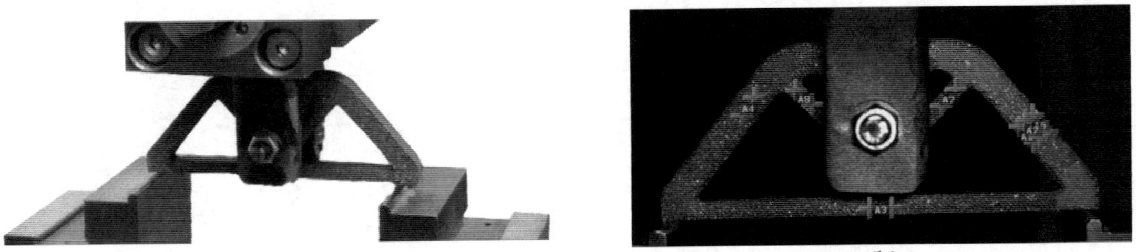

(a) (b)

Figure 3. The three-point bending arrangement of the reinforced geometry (a) and the camera image captured by the DIC system recording the exact deformations (b).

5. The simulation

5.1. CAD modelling of geometry and fiber reinforcement

For the implementation of BCS, the geometry of both the reinforcement structure and the matrix material were modelled at the micro-level, allowing tracking of the intralaminar interactions at the interfaces between the reinforcement and the matrix material. We used the simplification of merging the same layers in the construction direction (Z) of the specimen. In doing so, we sacrificed the modelling of interlaminar phenomena between the layers, but since the component was not dominated by such failure events, we did not lose any significant information by this simplification. The matrix material was modelled as a solid geometry, while the centerline of the fiber reinforcement was modelled with a surface strip of a width corresponding to the added thickness of the layers (Figure 4). Since the thickness of the printed layer is 0.125mm and filament reinforcement has a diameter of 0.375mm, the laid filament bundle can cover a width of ~0.89 mm in the XY plane.

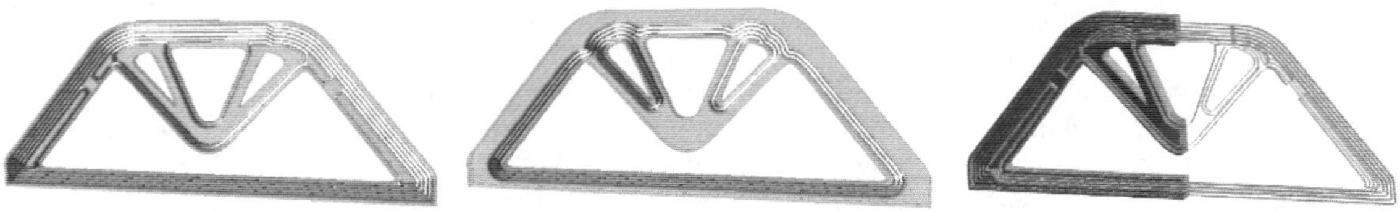

(a) (b) (c)

Figure 4. The merged geometry slices of the two applied reinforcement structures (a, b), from which the final VEM quarter model was constructed (c).

5.2. Simulation environment

The simulation was built in the Ansys Workbench 2023R2 environment. Here, we first separated the characteristic reinforcement structures (Figure 2), then solid bodies and the parts modelled as surfaces. The reinforcement structure was prepared in an ACP (Ansys Composite PrePost) environment while the matrix materials were pre-processed in the Static Structural module. Later, these modules were merged into a common Static Structural module where the final loading and contact boundary conditions were determined. Using the edges of the central surfaces representing the fiber reinforcement (Figure 5 a), we determined the trajectory in the ACP module for the implementation of BCS, on which we swept the coordinate system representing the directions of fiber reinforcement and assigned the local fiber directions to the surface elements (Figure 5). Then the central surface was symmetrically thickened according to actual fiber thickness, thus a VEM model of fiber reinforcement was created (Figure 5 c).

(a) (b) (c)

Figure 5. The steps of the Base Coordinate Sweep (BCS) are defining the reinforcing structure (a), performing the coordinate sweep where the green arrows indicate the direction of the fibers (b), determining the spatial extent of the reinforcing structure, and creating the finite element mesh (c).

The internal constraints of the characteristic sections (the connection of the connecting surfaces of the fiber reinforcement to themselves and the matrix) were determined in the next step, where the components were coupled by bonded constraints that are deactivated in the case of stresses above a limit value on the given surface, where shear and normal components were defined separately. With this method, separation phenomena between the printed components were modelled. The finished components were then merged into a common module, and the connections between them were also defined with a bonded constraint that terminates above the critical stress. The test specimen was defined by constraints corresponding to three-point bending with the introduction of vertical displacement through a pin passed through the center of the specimen and symmetrical support spaced at 80 mm. Both external constraints connected to the specimen with frictionless contacts.

The simulation aims to model progressive failure processes, therefore the use of explicit FEA methods is preferable, due to their ability to handle change in the substitutive properties during the calculation process far more stable, than their implicit counterparts. For that reason, we used the LS-Dyna solver integrated into Ansys Workbench.

Due to the complexity of the model and the existing symmetries, we worked with a quarter model, applying a frictionless planar constraint in the symmetry planes. We used a neglection—during printing the machine always had to cut the continuous fiber of the reinforcement structure somewhere, where reinforcement was consequently discontinuous. This is done automatically, always in a different location in the structure, and therefore the local weakening effect is always in a different place. For this reason, we simplified the reinforcement structure by assuming that it is continuous and there are no interruptions. Double symmetry can only be used with these simplifications. As a further simplification, the problem was not considered with the actual speed of load. The 5 mm crosshead displacement was applied to the system in 0.01 seconds. This was done to further reduce the computational cost of the problem. The 0.5 mm element size of the mesh and the density of the material resulted in a timestep of 4.5×10^{-5} ms, to which an adaptive mass scaling with a 1.7% increase in total mass was applied, to further reduce runtime. The mesh was constructed from linear hex8 and wed6 elements with an average quality of 0.9483±0.082 according to the built-in Ansys metric (1 being the maximum, at which point the mesh is perfect). With these settings, the simulation runtime at full processor use was ~2 hours per iteration on an Intel I7-7700K, 4.2 GHz processor with 4 cores and 8 logic threads.

5.3. *The applied simulation material model*

The MAT54 material card, also known as the "Enhanced composite damage" model, was selected for modelling the failure characteristics. The main parameters of the material card are shown in Figure 6 a), while its behavioral characteristics are shown in Figure 6 b). The material card can change the constitutive properties of the material after reaching certain threshold values and can model the behavior of the material by deleting elements declared as destroyed, causing so-called element death in the system. The MAT54 card utilizes the Chang-Chang failure criterion, described in more detail in the LS-Dyna material model manual [14]. After defining the constitutive, strength, and strain limit parameters and the failure criterion of the material model (Figure 6 a) groups 1, 4, 6, 7, respectively) and its local orientation parameters (Figure 6 a) group 2), the progressive failure parameters were defined and fine-tuned (Figure 6 a) groups 3, 5). The initial values of these were taken from the publication of Feraboli et al. [15] and were tuned in an iterative process to show the closest match with our test curves and with the local deformations obtained with DIC.

1. Constitutive properties: RO, EA, EB, EC, PRBA, PRCA, PRCB, GAB, GBC, GCA, KF	
2. Local material axes: AOPT, XP, YP, ZP, A1-A3, MANGLE, V1-V3, D1-D3	
3. Shear weighing factors: ALPH, BETA	**3. Deletion parameters:** DFAILM, DFAILS, TFAIL, DFAILT, DFAILC, EFS
5. Damage factors: SOFT, FBRT, YCFAC	
6. Material strengths: XC, XT, YC, YT, SC	**7. Failure criterion selection:** CRIT

(a)

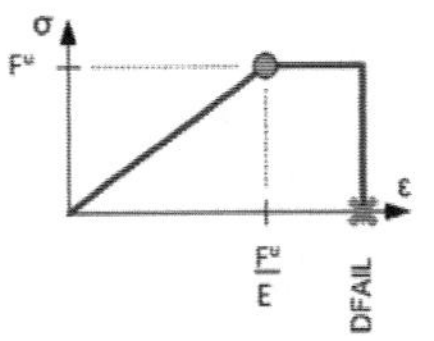

(b)

Figure 6. Table (a) [16] and graph (b) [17] show the behavior of the engineering constants of the MAT54 material card in logical groups. The purple dot indicates the location of ply failure, and the red X indicates the location of element deletion due to element death.

The different sections of the complex curves are dominated by the effects of the different engineering constants, so for example, the moduli can be tuned well from the initial linear section, and information on the sequence and limit values of the different failure modes can be obtained from the local force maxima at failure (Figure 7: numbered local failures). The difficulty of the task is due to the co-occurring effects, which are difficult to separate from each other and which, by their nature, cannot be tuned independently. This can be addressed either by performing several simpler tests separately or by parallel simulation validation on independent, complex problems.

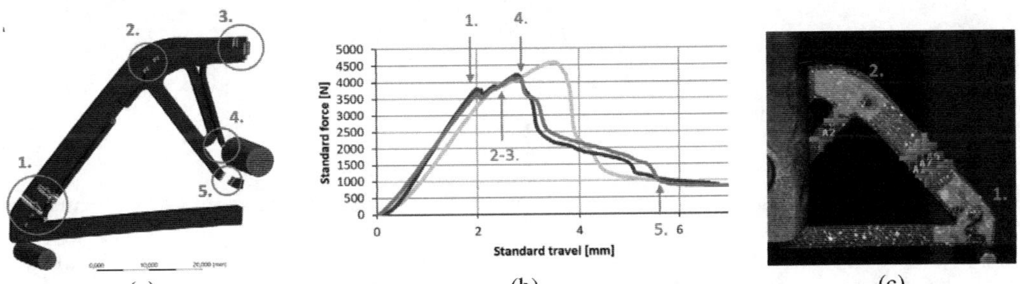

Figure 7. The location, time, and extent of failure of the elements (a) had to be adjusted based on the curves from the test and the video image of DIC strain measurement (c), in which the adjustment of the non-physical parameters was most challenging.

6. Results and their evaluation

In the fine-tuning of the parameters, the 45th iteration led to the simulation result shown in red in Figure 8, which was considered to be in good agreement with the tests. The abrupt drop in the simulation's reaction force values at 3.2 mm deformation, followed by a "rebound", is attributed to the breaking of the fiber bundles during the load, and the elastic energy release, which is not damped in the simulation, where the matrix material is modelled as linearly elastic. In reality, the PA6-based matrix material is a highly viscoelastic polymer, and its damping effect significantly influences the second, downward half of the failure curve.

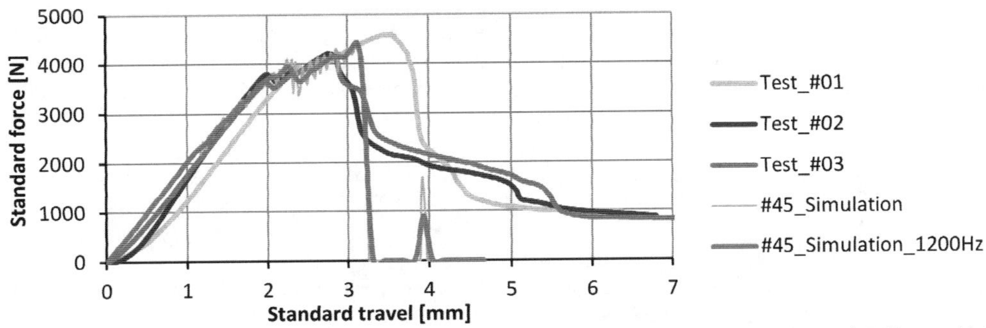

Figure 8. The measured force-deformation curves (blue) and the values of the final simulation (pink), on which a 1200 Hz Butterworth filter was applied (red) to improve the transparency of the values.

The values of the engineering constants for the last simulation can be found in Table 1 (b). The matrix material was modelled with isotropic material properties throughout, with a modulus of 1.4 GPa and a Poisson's ratio of 0.39. All bonded constraints in the final simulation, including delamination phenomena, were set with a maximum normal and shear load of 60 MPa. When this threshold was exceeded, the constraint was removed between the nodes of the respective elements.

Table 1. Material properties of the fiber-reinforced filament used for the initial simulation (a) and the tuned MAT54 material card (b). The constants in grey were unknown in the initial simulation and were determined during the simulation iterations, while the crossed-out parameters were not used in the final simulation.

*MAT_054 (ENHANCED_COMPOSITE_DAMAGE): Mpa, sec, mm, g/cm^3							BASELINE
mid	ro	ea	eb	ec	prba	prca	prcb
0	1,4	70650	1790	1790	0,008112	0,008112	0,41
gab	gbc	gca	kfail	aopt			
700	630	700	0	0			
xp	yp	zp	a1	a2	a3	mangle	
0	0	0	0	0	0	0	
v1	v2	v3	d1	d2	d3	dfailm	dfails
0	0	0	0	0	0	0,246	0,176
tfail	alph	soft	fbrt	ycfac	dfailt	dfailc	efs
0,00E+00	0	0	0	0	0,027	-0,022	0
xc	xt	yc	yt	sc	crit	beta	
-460	930	-36	36	32	0	0	

(a)

*MAT_054 (ENHANCED_COMPOSITE_DAMAGE): Mpa, sec, mm, g/cm^3							FINAL
mid	ro	ea	eb	ec	prba	prca	prcb
1	1,4	64700	1600	1600	0,008112	0,008112	0,41
gab	gbc	gca	kfail	~~aopt~~			
627	570	627	6470	0			
~~xp~~	~~yp~~	~~zp~~	~~a1~~	~~a2~~	~~a3~~	~~mangle~~	
0	0	0	0	0	0	0	
~~v1~~	~~v2~~	~~v3~~	~~d1~~	~~d2~~	~~d3~~	dfailm	~~dfails~~
0	0	0	0	0	0	0,25	0
tfail	alph	soft	fbrt	ycfac	dfailt	dfailc	efs
5,00E-10	0	0	0,8	3	0,027	-0,02	0
xc	xt	yc	yt	sc	crit	~~beta~~	
-573	860	-38	38	38	54	0	

(b)

7. Discussion

We presented a new method for the finite element modelling of continuous fiber-reinforced composites produced by automated manufacturing. We used the base-coordinate sweep method, coupled with the LS-Dyna explicit environment, and a MAT54 material card tailored for UD composites. The simulations and the validation tests were carried out on a topology-optimized test specimen subjected to three-point loading, which proved to be a good choice due to the complex load states in the specimen. The verification tests were performed with the addition of a DIC system, the results of which were later used to explore the local failure characteristics and finally compared with our simulation results. The constitutive parameters of the material model were partly taken from literature data and partly from our tests and calculations, while the unmeasured constants, which have only mathematical content, were determined in a heuristic, iterative way. The obtained results demonstrate a promising correlation with the measured values, but our investigation shows that the viscoelastic properties of the matrix material cannot be neglected, so their proper modelling requires further work. Overall, the presented method provides a suitable modelling basis for continuous fiber-reinforced composites produced by automated methods, thus greatly improving the engineering applicability of these structural materials produced with novel technology.

Acknowledgments

The project supported by the Doctoral Excellence Fellowship Programme (DCEP) is funded by the National Research Development and Innovation Fund of the Ministry of Culture and Innovation and the Budapest University of Technology and Economics, under a grant agreement with the National Research, Development and Innovation Office.

The research reported in this paper was supported by the National Research, Development, and Innovation Office (NRDI, Hungary) through grants OTKA FK134336 and OTKA K 138472.

Project no. TKP-6-6/PALY-2021 has been implemented with the support provided by the Ministry of Culture and Innovation of Hungary from the National Research, Development, and Innovation Fund, financed under the TKP2021-NVA funding scheme.

Project no. 2022-2.1.1-NL-2022-00012 „National Laboratory for Cooperative Technologies" has been implemented with the support provided by the Ministry of Culture and Innovation of Hungary from the

National Research, Development, and Innovation Fund, financed under the National Laboratories funding scheme.

References

[1] Sigmund O and Maute K 2013 Topology optimization approaches: A comparative review *Struct. Multidiscip. Optim.* **48** 1031–55
[2] Hou Z, Tian X, Zheng Z, Zhang J, Zhe L, Li D, Malakhov A V. and Polilov A N 2020 A constitutive model for 3D printed continuous fiber reinforced composite structures with variable fiber content *Compos. Part B Eng.* **189** Paper ID: 107893
[3] He Q, Wang H, Fu K and Ye L 2020 3D printed continuous CF/PA6 composites: Effect of microscopic voids on mechanical performance *Compos. Sci. Technol.* **191** Paper ID: 108077
[4] Somireddy M, Czekanski A and Atre S V. 2022 Modelling of Failure Behaviour of 3D-Printed Composite Parts *Appl. Sci.* **12** 1–13
[5] Oberlercher H, Heim R, Laux M, Berndt A, Becker C and Amancio- S T 2021 Additive manufacturing of continuous carbon fiber reinforced polyamide 6 : The effect of process parameters on the microstructure and mechanical properties *Procedia Struct. Integr.* **34** 111–20
[6] Chen Y and Ye L 2021 Topological design for 3D-printing of carbon fibre reinforced composite structural parts *Compos. Sci. Technol.* **204** Paper ID: 108644
[7] Szederkényi B 2021 *Újszerű tervezési és gyártási módszer alkalmazhatóságának vizsgálata a Formula Student versenyautó polimer kompozit alkatrészének fejlesztésében* (Master thesis, Budapest University of Technologies and Economics, Department of Mechanical Engineering)
[8] Sauer M 2018 *Evaluation of the Mechanical Properties of 3D Printed Carbon Fiber Composites* (South Dakota State University, Department of Mechanical Engineering)
[9] Zhang H, Chen J and Yang D 2021 Fibre misalignment and breakage in 3D printing of continuous carbon fibre reinforced thermoplastic composites *Addit. Manuf.* **38** Paper ID: 101775
[10] Goh G D, Toh W, Yap Y L, Ng T Y and Yeong W Y 2021 Additively manufactured continuous carbon fiber-reinforced thermoplastic for topology optimized unmanned aerial vehicle structures *Compos. Part B Eng.* **216** Paper ID: 108840
[11] Blok L G, Longana M L, Yu H and Woods B K S 2018 An investigation into 3D printing of fibre reinforced thermoplastic composites *Addit. Manuf.* **22** 176–86
[12] Markforged Material Datasheet Composites
[13] Ghebretinsae F 2019 *Mechanical testing and finite element analysis Of 3D printed continuous carbon fiber reinforced Onyx® thermoplastic* (Master thesis, Universitetet i Stavanger, Department of Engineering Structures and Materials Science,)
[14] LS-Dyna LS-DYNA Manual: Volume II R13
[15] Feraboli P, Wade B, Deleo F, Rassaian M, Higgins M and Byar A 2011 LS-DYNA MAT54 modeling of the axial crushing of a composite tape sinusoidal specimen *Compos. Part A Appl. Sci. Manuf.* **42** 1809–25
[16] Wade B, Feraboli P, Osborne M and Rassaian M 2015 *Simulating Laminated Composite Materials Using LS-DYNA Material Model MAT54 : Single-Element Investigation* (Seattle)
[17] Wade B, Feraboli P and Osborne M 2015 *Simulating Laminated Composites Using LS-DYNA Material Model MAT54 Part I : [0] and [90] Ply Single-Element Investigation* (Seattle)

Flow curve, necking and Finite Elements

L Tatár[1], P Bereczki[2]

[1] HUN-REN Centre for Energy Research, H-1121 Hungary, Budapest, Konkoly-Thege Miklós út 29-33.

[2] University of Dunaújváros, H-2400 Hungary, Dunaújváros, Táncsics M. u. 1/A

E-mail: bereczkip@uniduna.hu

Abstract. Reliable flowcurves are essential for many FE models. The flowcurves are usually determined via tensile tests. Simple formulas applied to raw data obtained from tensile tests can only be considered reliable until necking. After the onset of necking stress and strain fields become three-dimensional and the assumption of stress uniaxiality is not valid anymore. At the University of Dunaújváros experiments were carried out to reveal the real behaviour of specimens during tensile tests by using specimens with *rectangular* cross-section having regular grid marks. By post-processing the photos of the deforming specimens taken during tests, using the Choung-Cho correction method the approximation of post-necking behaviour has been improved considerably over the not corrected case.
Onset of the necking has been studied via FE models for *cylindrical* specimens using the previously determined flowcurve. At the onset of the necking the Considère criterion is equivalent to the independence of the axial force from cross-section surface. Based on this criterion we created theoretical "flowcurves" used by FE analyses, resulting in identification of typical behaviours.

1. Introduction

Many applications use FE modelling of plastic deformation. These cases fall into different categories, e.g. simulation of metal forming processes, simulation of accidental situations where large, or small plastic deformation occurs, checking limiting stresses for a structure, simulation of crack propagation in the elasto-plastic regime, etc. Depending on the needs, simulations can be done with different approaches, e.g. small deformation plasticity vs. large deformation plasticity, kinematic hardening vs. isotropic hardening, viscoplasticity vs. rate-independent plastic flow, etc [1], [2]. These models can be considered as different approximations of the deformation phenomena. Selection of an adequate model depends on practical needs. For this study we limit to large scale plasticity, time-independent plastic flow, and von Mises yielding criteria. Thorough this article true (logarithmic) strains and true stresses are considered. For simulations we only used isotropic material models, although a slight anisotropy for certain specimens has been observed experimentally.

Flowcurves are usually obtained from tensile tests using longitudinal type strain gauges. Thus, for tensile specimens without notch at the beginning of the deformation process a safe assumption is that the stress field is uniaxial and an uniaxial stress-strain curve is obtained by using simple formulas. During simulations in many cases there is a need to compute strains and stresses which are beyond the validity range of flowcurves obtained from simple uniaxial formulas. The reliability of the simulations is strongly dependent on the reliability of input data, so there is a tremendous need to improve the applicability of the flowcurves.

IOP Conf. Series: Materials Science and Engineering **1313** (2024) 012003 doi:10.1088/1757-899X/1313/1/012003

There are different approaches for estimating the post-necking behaviour of the specimens. For cylindrical specimens, a well-known method is the Bridgman correction [3]. Practical application of the Bridgman correction requires measuring the minimal diameter of the specimen and the radius of the osculating circle at the location of necking. Although the method is quite old and gives fairly good results, due to measurement difficulties it is still not widely used yet. A more generally applicable approach presented in [4] considers a lower and an upper bound for the unknown part of flowcurve and uses a weighted average method for successive estimation of the flowcurve. Based on parametric finite element simulations Choung and Cho [5] established an easy-to-use formula which requires only a few parameters, determined from tensile data.
The present article focuses on important side results obtained during trial-and-error steps taken during development of our optically assisted material testing method.

2. Obtaining the flowcurve and simulation of the tensile testing process

At the University of Dunaújváros we used specimens with square and rectangular cross-sections. Rectangular grid has been printed on surfaces of specimens. A two-camera system has been developed in-house to record the real shape of the specimens during deformation. Tracking the marks has been done by the Lucas-Kanade method [6]. Using a regular grid instead of irregular patterns and digital image correlation (DIC) (as presented in [7]) means that the output data is based on less a priori assumptions than those involving DIC. Thus, any irregularities of the deformation can be observed on photos (by naked eye) and on processed data. Unfortunately at large necking the reliability of tracking decreased considerably. Being work in progress, a complete evaluation of the patterns is yet to come.
For determining the flowcurve in the necking range the Choung-Cho method [5] has been used. For application of this method, from picture sets recorded optically we only used the contours of the specimens and the minimal cross-section determined from them. The steps of processing the graphical information are depicted schematically in Figure 1. The process is described in more detail in [8].

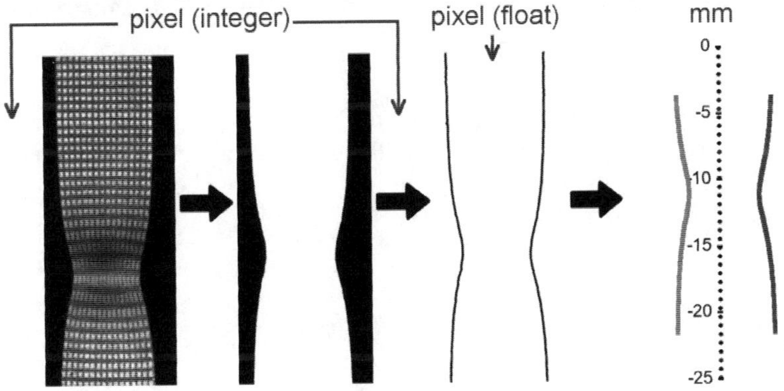

Figure 1. Steps taken during processing of graphical data

The raw (not corrected) and corrected flowcurves can be seen in Figure 2. Note, that the curve corrected with Choung-Cho method has a maximum point. This is caused by the polynomial nature of the correction. We consider, that the validity range of the corrected curve cannot extend beyond this maximum point.

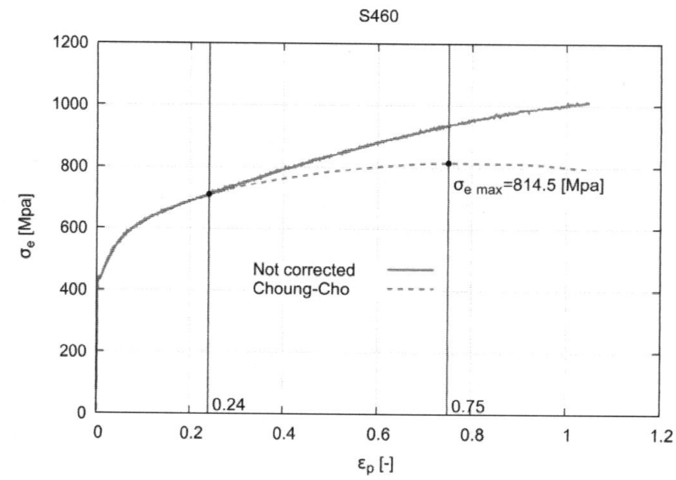

Figure 2. Original and corrected flowcurves

To check how accurate the corrected curve is, we simulated the tensile test process by FE. The mesh used for simulations can be seen in Figure 3. in its original and deformed state.

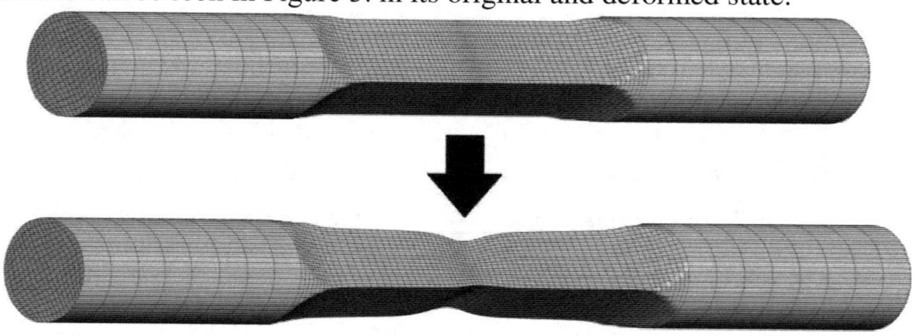

Figure 3. Mesh used for simulating the tensile test

The results show that by using the flowcurve obtained with the Choung-Cho correction method we obtain a good agreement with the experimental results on a larger interval than by using the not corrected flowcurve, however, the simulation is not precise at the end of the process. Figure 4 shows the measured and simulated Force – Thickness reduction diagrams at necking.

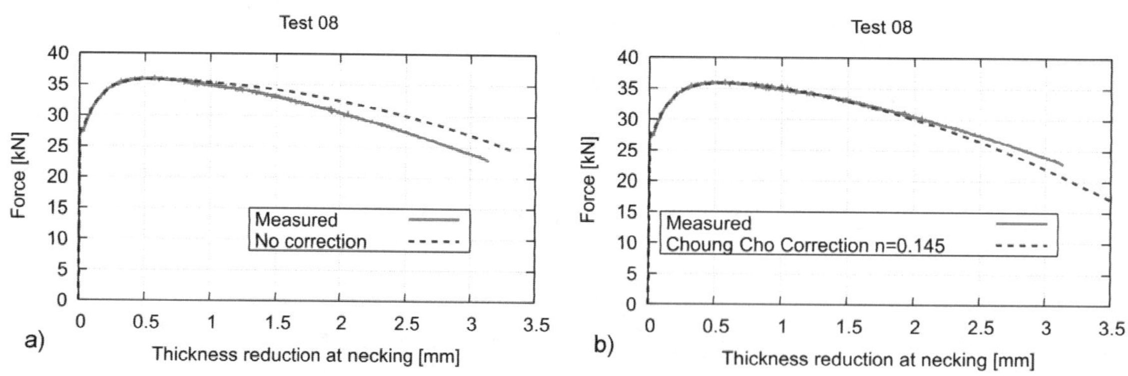

Figure 4. Simulation of the tensile test: (a) with original (not corrected) flowcurve, (b) with flowcurve corrected by the Choung-Cho method

3. Simulations of tensile tests for *cylindrical* specimens

Although the full 3D model gave a fairly good agreement with experimental results, and the experiments were carried out on specimens with *rectangular* cross-section, we continued our simulation work on *cylindrical* specimens. The reason for this is that full 3D simulations are very computationally expensive and for cylindrical specimens axisymmetric models can be used with a fraction of resources used. Besides, the vast majority of specimens used for tensile tests are cylindrical.
One of the main problems of optically assisted tensile tests is that we do not know beforehand the location of the necking. This means that even if we are interested only in the region of necking we have to look at the whole cylindrical part of the specimen, which leads to a considerable reduction of the practically achievable precision. To address this problem, we performed a series of axisymmetric FE simulations for specimens of different length. The specimens were characterized by their *L/D* ratio, where *L* denotes the length of the cylindrical part under deformation between the heads, *D* meaning its diameter. Subsequently in this article we denote *L* as "observed part". To minimize the effect of specimen heads on the results large radiuses were used in the transition zone (see Figure 5 for details). Note, that *L* is bigger than the gauge length as defined in standards.

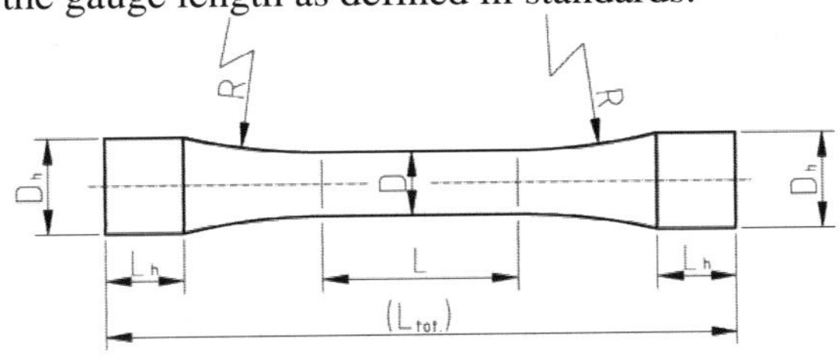

Figure 5. Specimens used for simulations

The *L/D* ratios followed the R10 Renard series [9], ranging from 1 to 10. This was complemented by the specimen with *L/D*=0, i. e. without cylindrical part. To eliminate the mesh effects on the results as much as possible the mesh contained square elements of the same size (except at the heads and their transition zones). Both specimen heads were included in the model. Diameter of the observed part was 8 [mm], elements were characterized by edge size of 0.2 [mm] for the observed part. Both heads were pulled axially in a concomitant way. The displacements were given to surfaces glued to the specimen heads (contact glue option for Hexagon Marc/Mentat FE software). Axisymmetric models were used. The mesh consisted of biquadratic, 8 nodes full integration elements. The material was homogenous, isotropic, with Young's modulus 210 000 [Mpa] and Poisson's ratio 0.3.
These simulations were not intended to be verified experimentally, they were used only for a parameter study. For this reason, the previously determined flowcurve has been used, but any kind of quasi-realistic flowcurve could have been used.
Every simulation has been performed in 500 steps. For every step we used the same displacement increments, but the displacement increments were different from analysis to analysis. The set-up is presented in Figure 6.

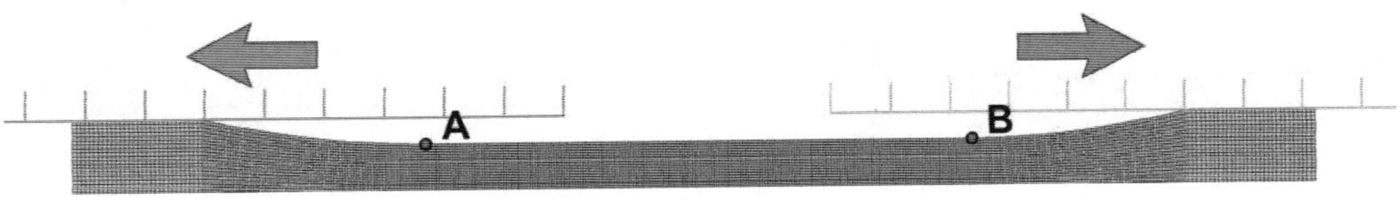

Figure 6. FE set-up for simulations

The displacement of each node for the observed part (between points A and B) has been collected from the result file for each increment, thus leading to sets of *simulated contour lines*. The precision of simulation (i.e. number of relevant digits) is much greater than that achievable with mechanical measurement devices.

3.1. Results of the simulations

The results show that only the specimen with *L/D*=0 (without cylindrical part) stays apart, the rest of the results are similar. Even a very short specimen with *L/D*=1 (with these extremely large radiuses at the necks) gives a difference for the uncorrected strain-stress curve resulting from simulations less than 1 % for the computed stresses at 0.3 logarithmic strain (computed from cross-section reduction) with respect to the specimen with *L/D*=10. Thus, using shorter specimens than those which are generally used might be acceptable. The results are presented in **Figure 7**.

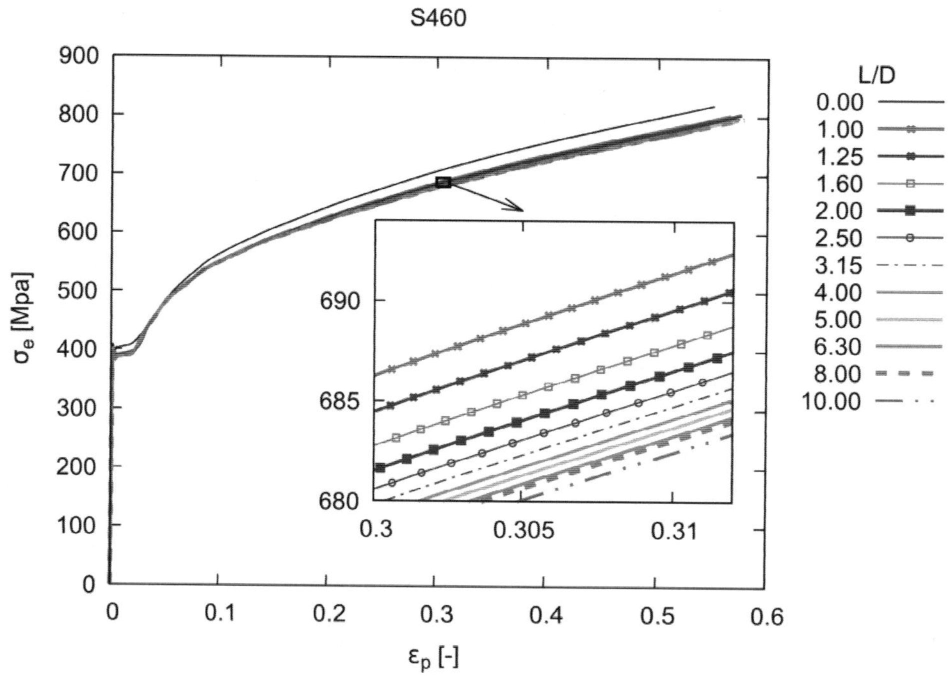

Figure 7. Plastic strain vs. axial stress curves from simulated specimens with different L/D ratios

On a more careful examination of the results one can observe, that even for an ideally homogenous material there are extremely small deviations from the ideal cylindrical shape caused by the combined effect of heads and the flowcurve. Our hypotheses is that these deviations will ultimately lead to macroscopically observable necking. This is presented in the following pictures (see notes).

Note 1. The combined effects of *flowcurve and idealized geometry* in their "pure" form are so small, that for realistic steel specimens and conventional instrumentation these effects would be (at least partially) covered by measurement noise, and effects of material and geometric inhomogeneities.

Note 2. For each picture different scales are used for axial (x) and radial (y) axes.

Note 3. The scale for *x* (axial) direction is different from case to case.

Note 4. The contour lines start from $y=4$ [mm] (8 [mm] diameter) and finish at about $y \approx 3$ [mm], which corresponds to 6 [mm] as the minimal diameter in the neck. The diagrams need to be read top-down as the process evolves. From the 500 increments of every simulation, we present every 50th increment on the diagrams. For comparison, see the specimens with *L/D*=1, *L/D=2.5*, *L/D=5* and *L/D*=10 (Figure 8 and Figure 9).

Although the global behaviour of the specimens (stress-strain obtained from force-diameter reduction) can be considered the same in each case, the localization of the necking for different L/D specimens follows a quite different path.

IOP Publishing
doi:10.1088/1757-899X/1313/1/012003

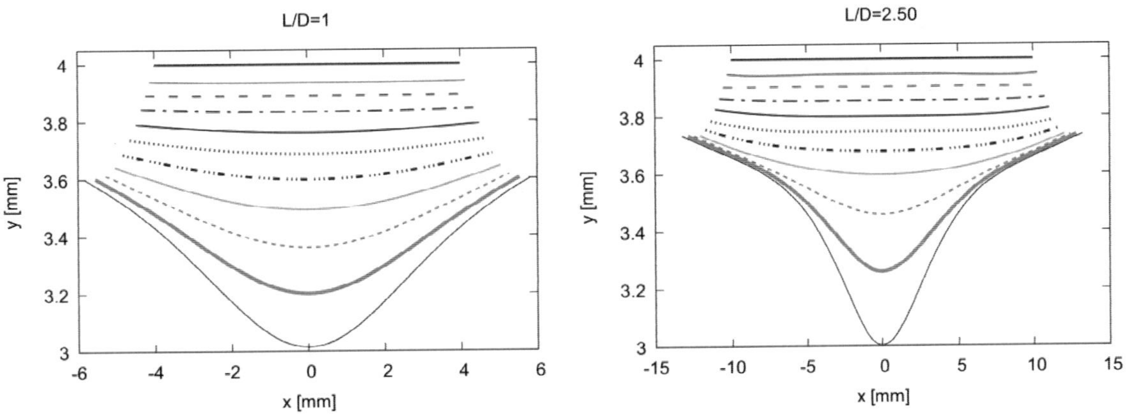

Figure 8. Evolution of the contourlines for the "observed part" for L/D=1 and L/D=2.5

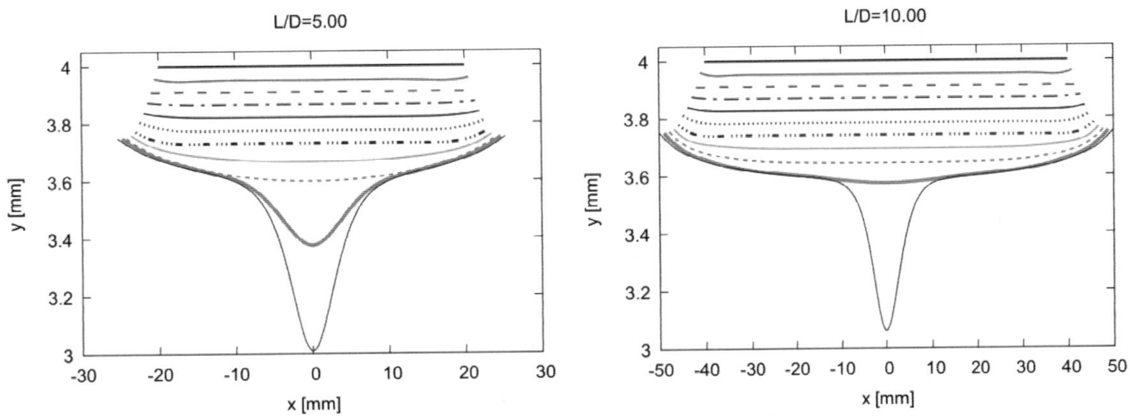

Figure 9. Evolution of the contourlines for the "observed part" for L/D=5 and L/D=10

For *L/D*=1 there is a "gentle" localization of the necking during the test, while in the case of *L/D*=10 there is no macroscopically measurable localization almost until the end of the process. Intermediate *L/D* ratios show intermediate behaviour.

Looking more into the detail of the localization process, one can observe, that even what looks simple from distance is not as simple at a scrutiny. This is demonstrated by a specimen with *L/D*=8, which is quite close to the *L/D* ratio of generally used specimens (the transition zone at the heads has usually a smaller radius for a specimen with the same diameter). The localization process is presented in Figure 10 (beware the difference in *y* scales for individual pictures!).

IOP Conf. Series: Materials Science and Engineering **1313** (2024) 012003 doi:10.1088/1757-899X/1313/1/012003

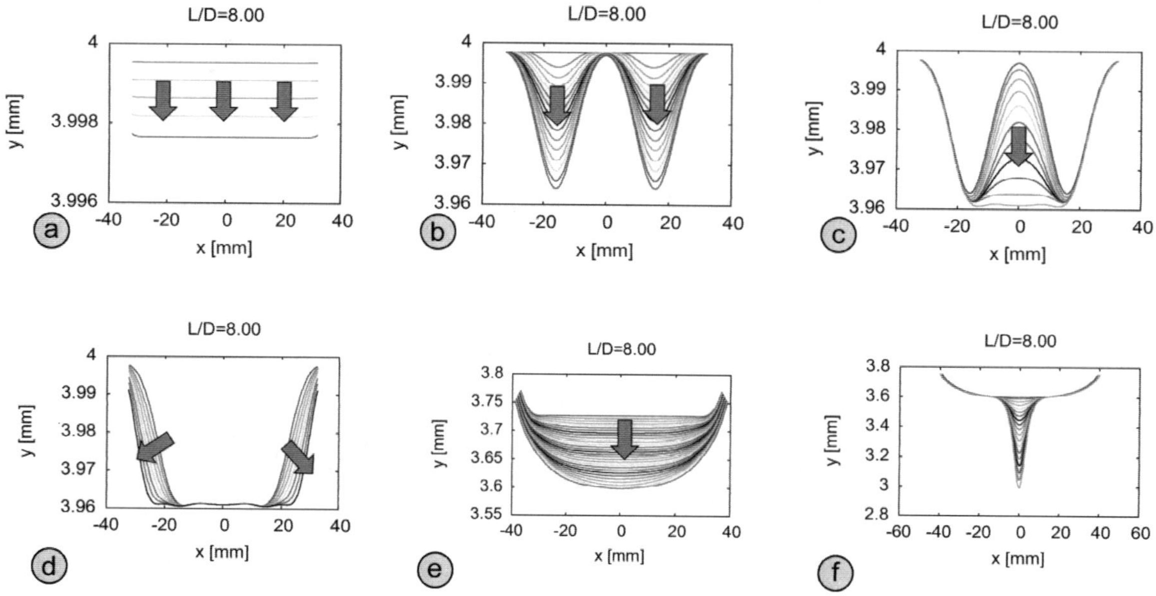

Figure 10. (a) uniform reduction, (b) localization, (c) delocalization, (d) more delocalization (e) dispersed necking (f) final localized necking

3.2. Theoretical considerations

Let us consider a specimen where necking begins (Figure 11). In static equilibrium the axial force is constant on the length of the specimen. We consider the limit state conform Considère's criterion [10] meaning that the force does not increase any more. **Note:** the criterion is applied to true strain – true stress, not engineering strain and stress as originally developed.

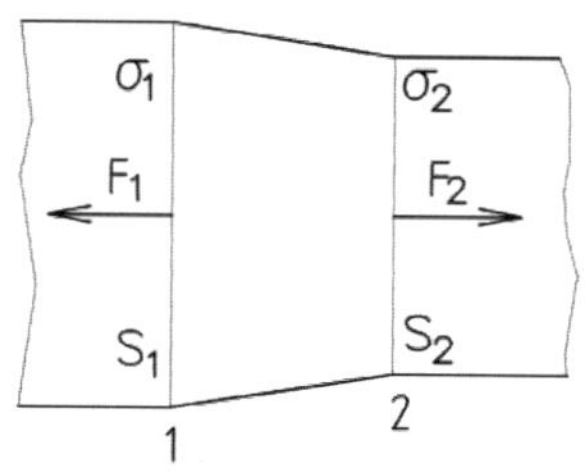

Figure 11. Forces, surfaces and stresses

From the constancy of the force: $F_1 = F_2$, which can be written for this case as: $S_1 \cdot \sigma_1 = S_2 \cdot \sigma_2$. In general, at the limit state $S \cdot \sigma = F$. In these formulas σ represents axial stress. For $S_1 \approx S_2$ $\sigma \approx \sigma_e$, where σ_e is the equivalent stress.

We can consider the stress as being a function of the cross section surface. $\sigma = \sigma(S)$. The $S \cdot \sigma = F$ equation is the equation of a hyperbola in the $\sigma - S$ coordinate system for $F = const$.

Figure 12 shows the $S - \sigma$ curve (derived from the $\varepsilon - \sigma$ curve of the material), and the $F = const$ hyperbolas. On this diagram S represents the minimal cross-section of the specimen perpendicular to its axis, while σ_{true} represents true (logarithmic) axial stress component derived from the axial force and the minimum cross-section. Reading of the diagram is backwards, in the direction of the curved arrow. Note, that at the right hand side the curve representing the material is more abrupt than the hyperbolas, on the middle range it is almost in-line, while at the left-hand side of the curve the hyperbolas are more abrupt. The yielding part on the diagram constitutes an exception from this general trend.

IOP Conf. Series: Materials Science and Engineering **1313** (2024) 012003
doi:10.1088/1757-899X/1313/1/012003

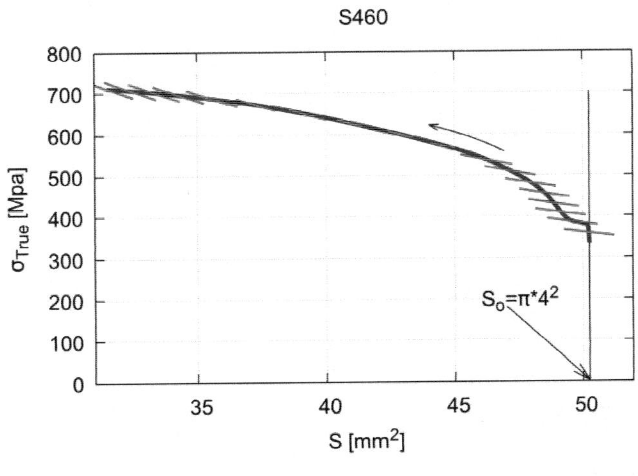

Figure 12. Curves derived from the flowcurve and the $F = const$ hyperbolas

On the other hand, from $\varepsilon = \ln\left(\frac{S_0}{S}\right)$ we have $S = S_0 \cdot e^{-\varepsilon}$, which leads to $\sigma = \frac{F}{S} = \frac{F}{S_0} \cdot e^{\varepsilon} = \sigma_0 \cdot e^{\varepsilon}$.

So, the hyperbolas in $\sigma = \sigma(S)$ correspond to the exponentials $\sigma = \sigma_0 \cdot e^{\varepsilon}$

3.3. Computations with artificial "flowcurves"
As pure theoretical cases, we consider three artificial "flowcurves".

"exponential"

$$\sigma = \sigma_0 \cdot e^{\varepsilon} \tag{1}$$

"hypoexponential", where a linear part is subtracted from the exponential:

$$\sigma = \sigma_0 \cdot e^{\varepsilon} - K \cdot \varepsilon \tag{2}$$

"hyperexponential", where a linear part is added to the exponential:

$$\sigma = \sigma_0 \cdot e^{\varepsilon} + K \cdot \varepsilon \tag{3}$$

The three "flowcurves" are depicted in Figure 13. for $\sigma_0 = 511$ [Mpa] and K=200 [Mpa])

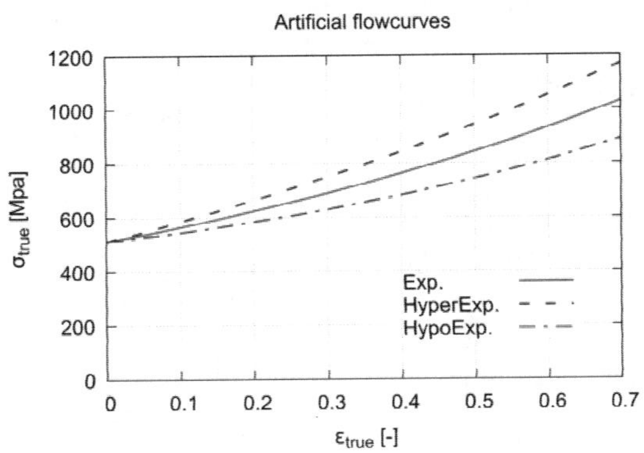

Figure 13. Artificial "flowcurves"

3.3.1. Simulation results using artificial flowcurves
The artificial flowcurves are applied to the specimens with L/D=10.

Figure 14. shows the results for the exponential flowcurve. Figure 15. shows the contour curves for hypo- and hyperexponential stress-strain curves.

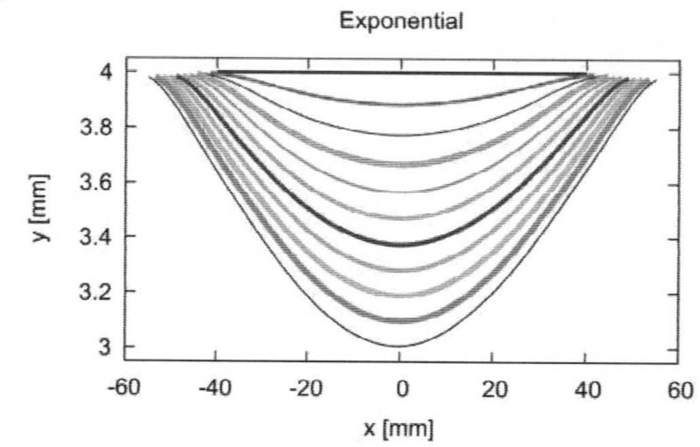

Figure 14. Contour curves for the exponential stress-strain curve

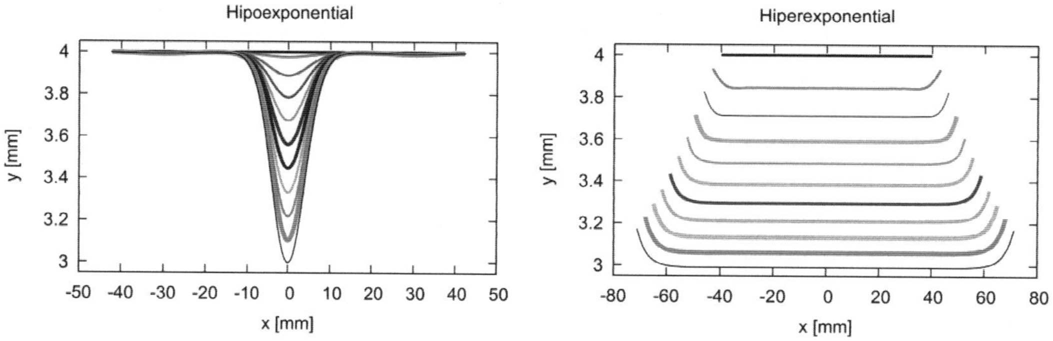

Figure 15. Countour curves for hypo- and hyperexponential stress-strain curves

It can be seen on the figures, that the three stress-strain curves lead to three different types of behaviour. In case of exponential curve there is a weak localization on the full length of the specimen. For hiperexponential curve there is only a transition zone from the specimen heads, otherwise there is no localization. For hypoexponential curve there is a strong localization, which begins at the very beginning of the testing process. Note, that the localization is numerically unstable, for different displacement increments we could even obtain two minimums (Figure 16). The curves presented in Figure 15. (left) were obtained for displacement increment of 0.025 [mm] while the curves in Figure 16. were obtained for displacement increment 0.019 [mm]. We consider, that the process of strong localization is not only numerically, but also physically unstable, thus any inhomogeneity (material, geometrical) can trigger localization. Fully developed twin necking as presented in Figure 16. is improbable for real specimens due to material and geometric inhomogeneities.

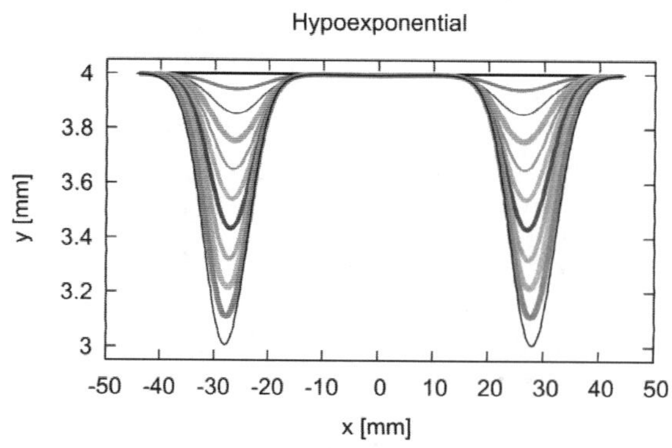

Figure 16. Contourcurves for hypoexponential stress-strain curve

The behaviours presented for contour curves obtained with artificial flowcurves can be considered as a reference for behaviours obtained from realistic flowcurve, taken into account that the realistic flowcurve has hyper- and hypoexponential parts with a large portion close to exponential. (see Figure 12. and Figure 10.)

4. Conclusions

Based on optical measurements and Choung-Cho correction we could increase the reliability range of flowcurves obtained for specimens with square cross-section. Using the previously obtained realistic flowcurve we made a series of simulations for cylindrical specimens with specimens with different *L/D* ratios. We can conclude, that although the global behaviour regarding diameter reduction vs. force is similar, the path leading to it is quite different. We could identify different basic behaviours.
Based on Considère's criterion we constructed artificial flowcurves, denominated exponential, hypoexponential and hyperexponential curves. These curves help us understand the basic behaviours identified for simulations based on realistic flowcurves.

5. Remaining work

The original set of experiments were not intended to address issues like transferability of results obtained from specimens with square cross-section to cylindrical ones, nor measuring necking with extremely high precision on cylindrical specimens. To separate the presented small effects of geometry and flowcurve from effects of material inhomogeneity and manufacturing precision we need very homogenous material and very high precision manufacturing. Designing a careful set of experiments on cylindrical specimens (taking into account the possibility of increasing optical measurement precision) will be considered for comparison with the FE computations.
A new set of simulations should include the radius at the neck too as a parameter for a parameter study. Using shorter specimens than usually accepted will be open to debate after having results of new simulations and experiments.

6. References

[1] Bruhns, O.T., 2020, *Large deformation plasticity*, Acta Mechanica Sinica, Volume 36, pp. 472-492
[2] Hexagon AB., 2023, *Marc® 2023.1, Volume A: Theory and user information*, pp. 65-108
[3] Bridgman, P. W., 1952, *Studies in Large Plastic Flow and Fracture*, (New York, McGraw-Hill), pp 9-16
[4] Ling, Y., 1996, *Uniaxial True Stress-Strain after necking*, AMP Journal of Technology Vol. 5 pp. 37-48.
[5] Choung, J., Cho, S.-R., 2008, *Study on true stress correction from tensile tests*. Journal of Mechanical Science and Technology. 22. pp. 1039-1051.
[6] Lucas, B. D., Kanade, T., 1981, *An iterative image registration technique with an application to stereo vision.* Proceedings of Imaging Understanding Workshop, pp. 121-130
[7] Tong, W., 2005, *An Evaluation of Digital Image Correlation Criteria for Strain Mapping Application,* Strain, Volume 41, pp. 167-175.
[8] Fekete, T., Antók, D., Tatár, L., Bereczki, P., 2022, *Extending reliability of FEM simulations, based on optically assisted tensile tests and digital twins*, Procedia Structural Integrity, Volume 42, pp 1684-1691
[9] ISO 3:1973 Preferred numbers. Series of preferred numbers.
[10] Considère, A., 1885. *Mémoire sur l'emploi du fer et de l'acier dans les constructions.* Annales des Ponts et Chaussées 9. (Paris, Ch. Dunod,), pp 574-605.

Acknowledgments

The measurements were done in the framework of the project no. 2020-4.1.1-TKP2020, implemented with the support provided by the Ministry of Innovation and Technology of Hungary from the National Research, Development and Innovation Fund, financed under the TKP 2020 funding scheme.

Tensile test and finite element simulation of coarse-grained thin aluminium sheets

A Szlancsik[1,*] **and D Varga**[1]

[1] Department of Materials Science and Engineering, Budapest University of Technology and Economics, Műegyetem rkp. 3., H-1111 Budapest, Hungary

*E-mail: szlancsik.attila@gpk.bme.hu

Abstract. The aim of this work was to investigate the ability of finite element simulations applied to specimens containing only a few grains for tensile testing. For this purpose, two different finite element models were created: one with individual grains with individual flow curves generated by viscoplastic self-consistent (VPSC) calculations and one with uniform material properties throughout the cross-section of the sample. In comparison of the two models with experimental tensile tests, the following conclusions can be yielded. The measurement and the simulation with individual grains have the same deformation shape. Also, the mechanical properties (yield strength, tensile strength and elongation at break) were within 10% error. However, if the measurement results were compared to the simulation with homogeneous material, it showed an error up to 53%. Therefore, it can be stated that the tensile tests of these types of specimens can be simulated, but only if the grains are modelled individually.

1. Introduction

Miniaturization, a significant trend in contemporary times, initially garnered notice within the nuclear industry by the late 1970s [1–3]. Due to the ongoing process of miniaturization, various fields like medical applications [4], microelectronics [5] and additive manufacturing [6,7] utilize thin films and sheets. In these instances, it has been well-documented that the mechanical properties undergo significant changes compared to those of the bulk material [8–10]. In mechanical loading scenarios, such as shaping, actuation, and fixation of such materials, the significance of free surface and size effects grows, also traditional continuum mechanics methods may diminish in validity under these circumstances [11]. Additionally, if the thickness of the component contains only a few grains, variations in the response to applied forces between parts can be substantial. Consequently, maintaining consistency in mechanical performance can prove challenging. Disregarding these effects in industrial design may lead to unforeseen issues, such as incorrect or too expensive solutions for small components. Many well-known phenomena in forming technology, such as plastic failure, contraction, texture formation, etc., require reconsideration as the ratio between the component and grain size decreases significantly [12]. In theory, the mechanical properties of micro components are influenced by four parameters: grain size, the quantity of grains within the specimen's thickness, specimen thickness, and the existence of free surfaces. When the grain size exceeds a few micrometres, reducing it generally enhances strength, a relationship commonly described by the Hall-Petch relationship [13,14].

Due to the difficulty of measuring miniaturised samples, new measurement methods have been developed. This resulted in the development of miniature specimens of different sizes of tensile specimens since the dimensions of the specimens are not defined in a standard [15]; therefore, the dimensions are determined according to the need and availability of the materials. A new measuring method has also been developed, namely the small punch test [16,17], which can be used to determine the fracture strength [18] and mechanical properties [19,20] (yield strength and tensile strength) of the given specimen. The inverse finite element method is employed for determining the behaviour of materials under uniaxial tensile testing, specifically the true stress vs. true plastic strain curve. Husain et al. [21] developed an inverse finite element method, with the help of which the small punch test results give an adequate prediction of the true stress-strain behaviour and become comparable with the results of the standard tensile test.

Kumar et al. [22] simulated miniaturised tensile specimens using 2-D finite element analysis. The material parameters used were determined from a standard uniaxial tensile test. The results of a standard specimen and a miniaturised specimen were compared, and it was found that the value of the tensile strength was 2-4.5% lower in the case of the miniaturised specimens. Fülöp et al. [23] investigated ultrathin metal sheets experimentally and through simulation. The effect of the surface grains was examined in relation to the total grain number. In their finite element simulation, they used the single crystal approach for multi-crystal analysis, and their goal was to analyse ultrathin sheets based on crystal plasticity and to verify the predictive ability of the applied finite element simulation. The results showed that the yield stress depends on the number of grains in the thickness direction, and due to the different crystal orientations, the deformation was not uniform under homogeneous tensile stress.

The goal of Kumar et al. [24] was to determine the minimum amount of material required for the properties of the miniaturised specimen to be comparable to that of the standard specimen. For this, finite element simulation was also used, and it was found that the yield stress, tensile strength and elongation values stabilised for a thickness of 0.25 mm, which corresponds to a thickness/grain size ratio of 6, 3.4 and 1.5 in the case of 20MnMoNi55, CrMoV steel and SS304LN materials.

It can be seen from the previous articles that many different approaches have been tried to determine the relationships between bulk materials and materials with only a few grains. However, no procedure has yet been created that could reliably establish this relationship, so the first step in our work is to examine whether the finite element simulation is suitable for replacing cumbersome and in many ways difficult to perform or reproduce measurements. To this end, we first carry out tests that can be reliably reproduced, and then use these to validate our finite element model. Using this model, even the complicated load conditions of the built-in components can be tested in the future.

2. Materials and methods

2.1. Experimental methods

In our experimental work, we used 1 mm thick and 10 mm wide Al99.8 strips. As a first step, the strips were annealed at 570°C for 30 min and then cooled in water. Then, the strips were subjected to an 12% deformation using an Instron 5965 universal material testing machine with uniaxial tension. The previously annealed and cold formed strips were heat treated again to induce recrystallization at 570°C for 30 minutes and then cooled in water. After heat treatment, 10 specimens with a width of 6 mm and a gauge length of 50 mm were produced by machining. The microstructure of the specimens was visualised by etching. The composition of the aluminium macro-etching solution is listed in Table 1.

Table 1. Composition of the aluminium macro-etching solution

Component	Distilled water	Nitric acid (65%)	Hydrofluoric acid (40%)	Hydrochloric acid (37%)
Ratio (Volume %)	18	16	16	50

After etching, the formed grain structure was examined with an Olympus SZX16 stereomicroscope in order to reproduce it by finite element simulation (Figure 1. (a)). The resulting specimens were subjected to tensile tests on an Instron 5965 universal material testing machine at a crosshead speed of 3 mm/min at room temperature. The electromechanical testing machine continuously recorded the load (F) and displacement (Δl) values during the test. These values were then plotted on a diagram in order to evaluate the measurements. First, the engineering stresses and engineering strains were calculated from the measured data, and then the yield strength ($R_{p0.2}$), tensile strength (R_m) and elongation at break (A) were determined.

2.2. Simulation
In order to simulate the tensile test of the specimens, a model was established using MSC Marc Mentat finite element analysis software. The specimen geometry was identical to the measured one, and the tension was applied with two rigid plates at the end of the specimen with glued contact. One plate was fixed, while the other moved with 3 mm/min. The 3-D model contained 11708 hexahedral elements, with an average side length of 0.41 mm and four elements in the thickness direction. Two different simulations were created. In the first one, the individual grains were selected, and four different material types were assigned to them randomly in order to simulate the specimens which were investigated previously (Figure 1.).

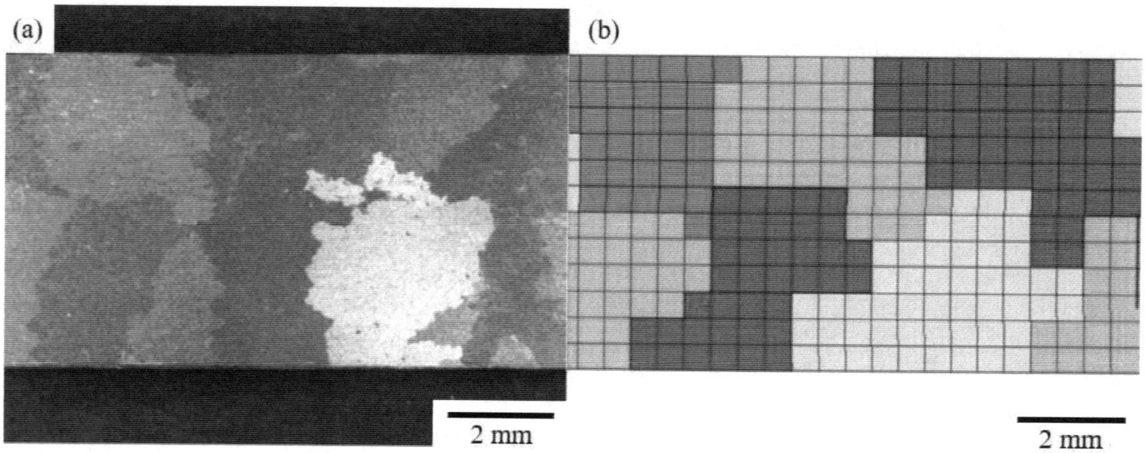

Figure 1. (a) Stereomicroscopic picture of the specimen after etching, (b) the individual grains selected in the finite element model. The colours represents the orientations as follows: red – orientation I, yellow – orientation II, green – orientation III and blue – orientation IV

The material properties in this case were obtained from a viscoplastic self-consistent (VPSC) calculation [25]. In order to get the flow curve of a single grain, the deformation field was given for the software, and only one element was defined in it. The Voce hardening law was used with the material properties given in Table 2. The four different flow curves were created with four different orientations of a single crystal. The investigated orientations were given by Bunge angles in degrees, which were as follows: (I) 0-0-0, (II) 10-0-0, (III) 20-0-0 and (IV) 45-0-0.

Table 2. Material properties of the Al99.8 used in the VPSC simulation

Elastic stiffness (GPa)	E11, E22, E33		E12, E13, E23	E44, E55, E66
	106.75		60.41	28.34
Voce hardening parameters (MPa)	$\tau 0$	$\tau 1$	$\theta 0$	$\theta 1$
	7.5	10.5	220.0	16.5

The other simulation assumed homogeneous material in the entire cross-section of the specimen. The flow curve of the material was given by equation (1)

$$\sigma = 22.2 + 112 \cdot \varphi^{0.416} \tag{1}$$

where σ – is the flow stress in MPa and φ – is the plastic strain. In the case of both simulations, the Young's modulus was 70 GPa, and the Poisson ratio was 0.33. Also, the Cockroft-Latham damage model was used with a 0.43 element removal threshold.

3. Results

In order to compare the results from the tensile tests and the simulations, the data points have been plotted in a common diagram (Figure 2.). It can be seen that there is a large difference between the two simulations. It shows the necessity to model the individual grains and to take into account the different deformations in those grains. Not only is the stress – strain curve higher than the measurement, but it can withstand more than twice the deformation before failure. However, with similar damage parameters, the model which contained the grains individually breaks in close alignment with the measured samples. Figure 3. depicts the total equivalent plastic strains at 12% engineering strain in case of the simulations and also a picture of a sample during measurement. The difference between the two types of models is significant. If the grains are neglected, the total equivalent plastic strain is homogeneous within the sample, which is to be expected since this is the reason why the specimens are designed in this way. However, if the grain size and the sample size are close to each other, the deformation becomes inhomogeneous. This inhomogeneity causes some grains to deform more, thereby reaching the critical deformation and breaking earlier. Also, the edges of the measured specimen are uneven, and this unevenness is also well represented by the simulation containing individual grains.

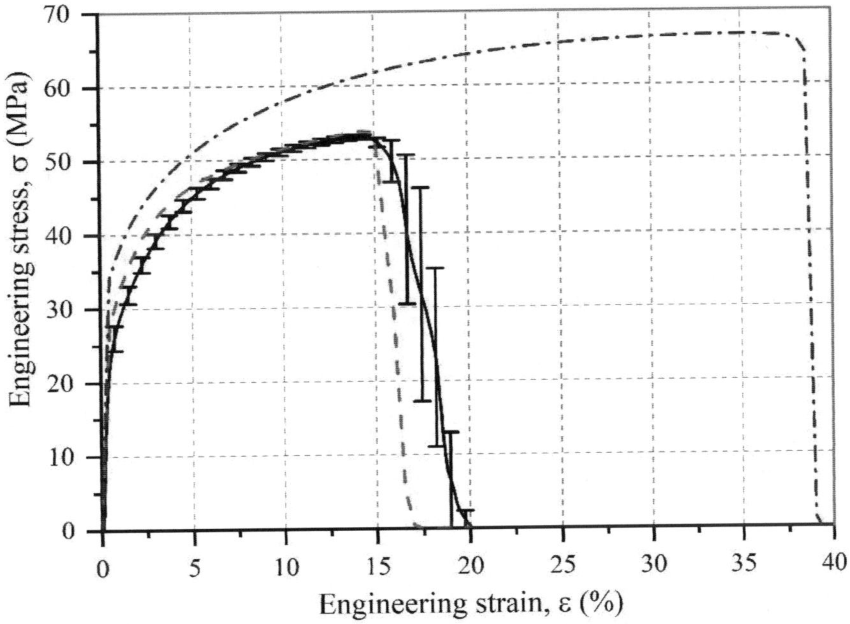

Figure 2. The engineering stress – engineering strain curves for the measurement (black – continuous line), simulation with individual grains (red – dashed line) and simulation with homogeneous materials (blue – dashed-dotted line).

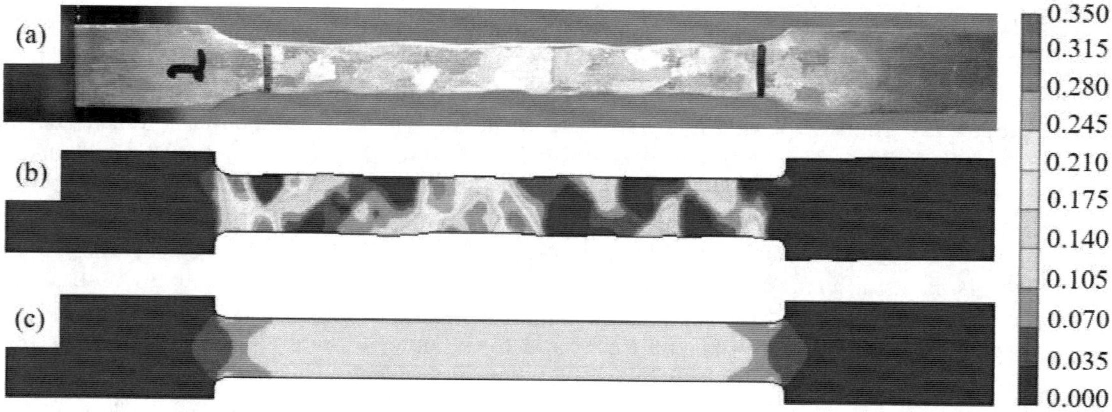

Figure 3. (a) Sample 2 during the tensile testing, (b) simulation with individual grains at 12% engineering strain, (c) simulation with homogeneous material at 12% engineering strain. In case of the simulations, the colours represent the total equivalent plastic strains.

Table 3. contains the mechanical properties determined with measurement and simulation. The percentage error between the mean measured value and the simulation results has been calculated. It can be seen that the simulation with individual grains is in good agreement with the measurement; only the elongation error is slightly bigger than 10%. The results from the simulation without grains are not in good agreement with the measurement results; namely, the elongation error exceeds 53% which is huge. The yield strength and the tensile strength are also above 20% error, which is unacceptable in this case.

Table 3. Mechanical properties obtained from measurements and simulations. The percentage error was calculated between the mean value of the measurements and the simulations.

	Measurement	Simulation (individual grains)	Error	Simulation (homogeneous material)	Error
$R_{p0,2}$ (MPa)	27.0 ± 0.9	29.9	9.7 ± 3.0 %	35.9	24.8 ± 2.5 %
R_m (MPa)	52.7 ± 0.6	53.8	2.0 ± 1.1 %	66.7	21.0 ± 0.9 %
A (%)	18.2 ± 1.3	16.5	10.3 ± 7.9 %	38.8	53.1 ± 3.4 %

4. Conclusion

The aim of our work was to investigate the capabilities of the applied finite element simulation of the tensile test of specimens containing only a few grains. For this purpose, two different finite element model was created: one with individual grains with individual flow curves, which were created by viscoplastic self-consistent (VPSC) calculations and one with homogeneous material properties in the entire cross-section of the specimen. The two models were compared with experimental tensile tests, and the following conclusions were drawn:

- The deformed shape of the specimen in the measurement and in the case of the simulation model with individual grains was similar.
- The specimens with few grains have different behaviour compared to the polycrystalline specimens, and their tensile tests cannot be simulated assuming bulk, homogeneous material.
- In case of the simulation with individual grains, the mechanical properties (yield strength, tensile strength and elongation at break) were within 10% difference, whereas when simulated as a homogeneous material, the difference was as high as 73%.
- It was shown that the individual grain simulation model is suitable for the simulation of tensile tests of specimens with few grains.

Acknowledgments
Project no. TKP-6-6/PALY-2021 has been implemented with the support provided by the Ministry of Culture and Innovation of Hungary from the National Research, Development and Innovation Fund, financed under the TKP2021-NVA funding scheme. This research was partly supported by the János Bolyai Research Scholarship of the Hungarian Academy of Sciences No. BO/00442/22/6.

References
[1] Karthik V, Kasiviswanathan K V. and Raj B 2016 *Miniaturised testing of engineering materials*
[2] Panayotou N, Atkin S, Puigh R and Chin B Design and Use of Nonstandard Tensile Specimens for Irradiated Materials Testing *The Use of Small-Scale Specimens for Testing Irradiated Material* (100 Barr Harbor Drive, PO Box C700, West Conshohocken, PA 19428-2959: ASTM International) pp 201-201–18
[3] Rosinski S, Kumar A, Cannon N and Hamilton M Application of Subsize Specimens in Nuclear Plant Life Extension *Small Specimen Test Techniques Applied to Nuclear Reactor Vessel Thermal Annealing and Plant Life Extension* (100 Barr Harbor Drive, PO Box C700, West Conshohocken, PA 19428-2959: ASTM International) pp 405-405–12
[4] Kirsten S, Schubert M, Braunschweig M, Woldt G, Voitsekhivska T and Wolter K J 2014 Biocompatible packaging for implantable miniaturised pressure sensor device used for stent grafts: Concept and choice of materials *Proc. 16th Electron. Packag. Technol. Conf. EPTC 2014* 719–24
[5] Zimprich P, Saeed U, Betzwar-Kotas A, Weiss B and Ipser H 2008 Mechanical size effects in miniaturised lead-free solder joints *J. Electron. Mater.* **37** 102–9
[6] Renkó J B, Kemény D M, Nyiro J and Kovács D 2019 Comparison of cooling simulations of injection moulding tools created with cutting machining and additive manufacturing *Mater. Today Proc.* **12** 462–9
[7] Kemény D M and Kovács D 2022 The Effect of Welding Parameters on the Corrosion Resistance of Austenitic Stainless Steel *Period. Polytech. Mech. Eng.* **66** 151–7
[8] Keller C, Hug E and Feaugas X 2011 Microstructural size effects on mechanical properties of high purity nickel *Int. J. Plast.* **27** 635–54
[9] Sergueeva A V., Zhou J, Meacham B E and Branagan D J 2009 Gage length and sample size effect on measured properties during tensile testing *Mater. Sci. Eng. A* **526** 79–83
[10] Dedov A and Klevtsov I 2012 Comparison of direct and indirect methods of tensile properties determination for post-exposed power plant steels *23rd DAAAM Int. Symp. Intell. Manuf. Autom. 2012* **1** 95–8
[11] Kohno Y, Kohyama A, Hamilton M L, Hirose T, Katoh Y and Garner F A 2000 Specimen size effects on the tensile properties of JPCA and JFMS *J. Nucl. Mater.* **283–287** 1014–7
[12] Janssen P J M, de Keijser T H and Geers M G D 2006 An experimental assessment of grain size effects in the uniaxial straining of thin Al sheet with a few grains across the thickness *Mater. Sci. Eng. A* **419** 238–48
[13] Hall E O 1951 The deformation and ageing of mild steel: III Discussion of results *Proc. Phys. Soc. Sect. B* **64** 747–53
[14] Petch N J 1953 The Cleavage Strength of Polycrystals *J. Iron Steel Inst.* **174** 25–8
[15] Zheng P, Chen R, Liu H, Chen J, Zhang Z, Liu X and Shen Y 2020 On the standards and practices for miniaturised tensile test – A review *Fusion Eng. Des.* **161**
[16] Manahan M P, Argon A S and Harling O K 1981 The development of a miniaturised disk bend test for the determination of postirradiation mechanical properties *J. Nucl. Mater.* **104** 1545–50
[17] Dyson C C, Sun W, Hyde C J, Brett S J and Hyde T H 2016 Use of small specimen creep data in component life management: a review *Mater. Sci. Technol. (United Kingdom)* **32** 1567–81
[18] Petersen D, Foulds J, Woytowitz P, Parnell T and Jewett C 1995 Fracture Toughness by Small Punch Testing *J. Test. Eval.* **23** 3
[19] Mao X, Takahashi H and Kodaira T 1991 Estimation of mechanical properties of irradiated nuclear pressure vessel steel by use of subsized CT specimen and small punch specimen *Scr.*

IOP Conf. Series: Materials Science and Engineering **1313** (2024) 012004 doi:10.1088/1757-899X/1313/1/012004

Metall. Mater. **25** 2487–90
[20] Dobeš F and Milička K 2002 On the Monkman–Grant relation for small punch test data *Mater. Sci. Eng. A* **336** 245–8
[21] Husain A, Sehgal D K and Pandey R K 2004 An inverse finite element procedure for the determination of constitutive tensile behavior of materials using miniature specimen *Comput. Mater. Sci.* **31** 84–92
[22] Kumar K, Pooleery A, Madhusoodanan K, Singh R N, Chakravartty J K, Dutta B K and Sinha R K 2014 Use of Miniature Tensile Specimen for Measurement of Mechanical Properties *Procedia Eng.* **86** 899–909
[23] Fülöp T, Brekelmans W A M and Geers M G D 2006 Size effects from grain statistics in ultra-thin metal sheets *J. Mater. Process. Technol.* **174** 233–8
[24] Kumar K, Pooleery A, Madhusoodanan K, Singh R N, Chatterjee A, Dutta B K and Sinha R K 2016 Optimisation of thickness of miniature tensile specimens for evaluation of mechanical properties *Mater. Sci. Eng. A* **675** 32–43
[25] Lebensohn R A and Tomé C N 1994 A self-consistent viscoplastic model: prediction of rolling textures of anisotropic polycrystals *Mater. Sci. Eng. A* **175** 71–82

Unidirectional fibre reinforcement of syntactic metal foams

B Szovák[1,2,*], J E Maróti[1,2], A Kemény[1,2] and I N Orbulov[1,2]

[1]Department of Materials Science and Engineering, Faculty of Mechanical Engineering, Budapest University of Technology and Economics, Műegyetem rkp. 3., H-1111 Budapest, Hungary
[2]MTA–BME Lendület „Momentum" High-performance Composite Metal Foams Research Group, Műegyetem rkp. 3., H-1111 Budapest, Hungary
[*]Corresponding author: szovak.benedek@edu.bme.hu

Abstract. Metal matrix syntactic foams are porous structures where the porosity is formed by introducing a porous filler material into the metal matrix. They have low density and absorb significant energy during compression but cannot withstand tensile stress. The main goal of this study was to apply unidirectional fibre reinforcement to syntactic metal foams. Therefore, a lightweight double-composite structure that can be utilised against tensile load and has advantageous properties during compression was created. Copper-nickel coated carbon fibre bundles were used as unidirectional long fibre reinforcement, lightweight expanded clay aggregate particles as filler material and Al99.5 aluminium as matrix material. The samples were created with vacuum infiltration and formed into cylindrical tensile specimens. Quasi-static tensile tests were carried out on the specimens and fibre trusses, and quasi-static compression was applied on specimens in the parallel and perpendicular direction to the unidirectional fibre truss.

1. Introduction

Industrial requirements for materials can conflict with each other, such as the expectation for lower weight and higher strength and toughness at the same time. While a single type of material cannot fulfil this expectation independently, a composite structure, where two (or more) advantageous material characteristics are paired up, can achieve the desired improvement. Examples of this are the metal matrix composites (MMCs). The application of different types of reinforcing material increases the mechanical properties of the metal matrix. The reinforcement can be in the form of ceramic particles [1–4] or different fibre structures [5–8]. The unidirectional fibre reinforcement can increase a structure's Young's modulus, tensile strength and elongation [9,10].

One of the option is to reduce the weight by using porous materials like metal foams [11–13]. Metal foams have low density and advantageous properties against compression. Based on the cell structure, they can be categorised as open-cell [14,15] and closed-cell metal foams [16]. Syntactic or composite metal foams are specific types of closed-cell metal foams where the porous structure is formed by a porous filler material. The size and material of this filler can significantly influence the properties of the final structure [17–19].

The goal of this study is to create a double-composite structure where the composite metal foam matrix is reinforced with unidirectional carbon fibre trusses. By that, to create a structure where the carbon fibre reinforcement can handle tensile stress while the metal foam matrix has a low density and advantageous properties against compression.

2. Materials and methods

2.1. Materials

Al99.5 aluminium was chosen as the matrix material owing to its high ductility. Lightweight expanded clay aggregate particles (LECAPs) were used as the filler material because they have low density and are easily machinable. Copper-nickel coated carbon fibre bundles (CFBs) were used as reinforcement. The coating serves the wetting; it helps to create a better bond between the matrix and the fibres. Macrophotos of the filler material particles and the CFBs are shown in Fig 1.

Figure 1.(a.) LECAPs filler material and (b.) cooper-nickel coated carbon fibre

2.2. Manufacturing

The CFBs were oriented with a tension frame. The frame was made of S235JR steel, and a 1.4301 (X5CrNi18-10) stainless steel net was attached at the ends. The CFBs were attached to these nets. Two different types of samples were prepared, one with and one without the twisting of the CFBs, to determine if the density of the CFBs had any effect. The frame was filled with the filler material while placed into the crucible. Extra filler material was placed on top of the frame, and the installation was fixed with another steel net with a finer hole size. An alumina quilt was placed as the final layer to clear out possible pollution from the molten matrix during infiltration. The tension frame (partially filled with CFBs) and the schematic cross-section of the infiltration installation are shown in Fig 2.

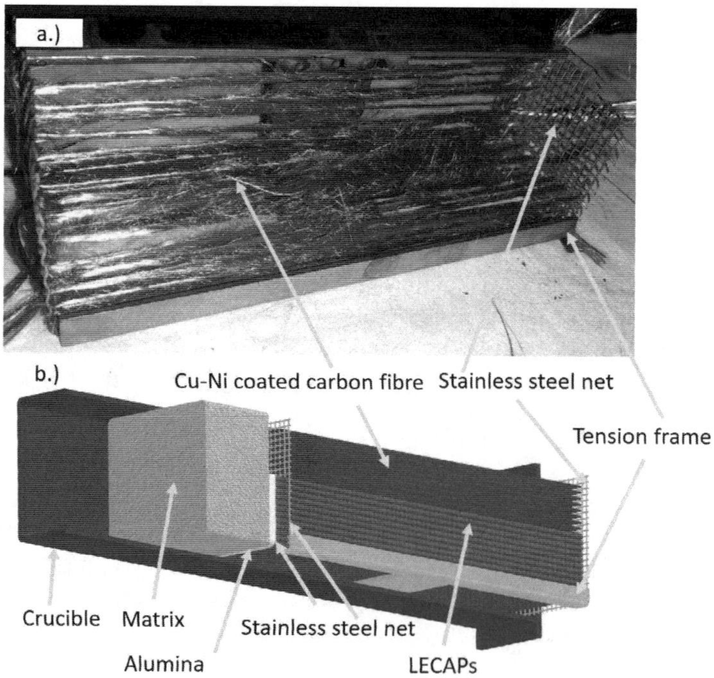

Figure 2.(a.) Tension frame and (b.) the schematic cross-section of the infiltration installation

The matrix material block was placed on top of the installation, and the filled mould was placed into the vacuum furnace, where it was pre-heated to 550°C and kept at that temperature for an hour while the vacuum built up. After the pressure reached 10^{-7} MPa, the temperature was raised to 750°C to melt the matrix, kept for an hour, and the infiltration was carried out with argon gas with a pressure of 0,4 MPa.

The samples were cut and machined into 5 pieces of Ø20×100 mm cylindrical tensile specimens, 13 pieces of Ø20×30 mm cylindrical and 11 pieces of 14×14×14 mm cubic compression specimens.

2.3. Methods

The mechanical properties of the specimens were examined with quasi-static tensile tests based on ISO 6892-1:2019 [20] and compression tests according to ISO 13314:2011 [21] standards. The cylindrical specimens were compressed in parallel, and the cubical ones were compressed perpendicularly to the direction of the reinforcement. The characteristic behaviour of the specimens during compression and tensile test is shown in Fig 3. The cylindrical specimens have the specific ductile behaviour of metal foams (due to the high ductility of the applied matrix material) (Fig. 3a.)), while the strain of the cubic specimens was hindered in one direction by the CFBs inside (Fig. 3b.)). The tensile specimens showed brittle behaviour with the appearance of the characteristic ~45° crack and little to no contraction (Fig. 3c.)) because of the brittle nature of the filler material and the reinforcing fibres.

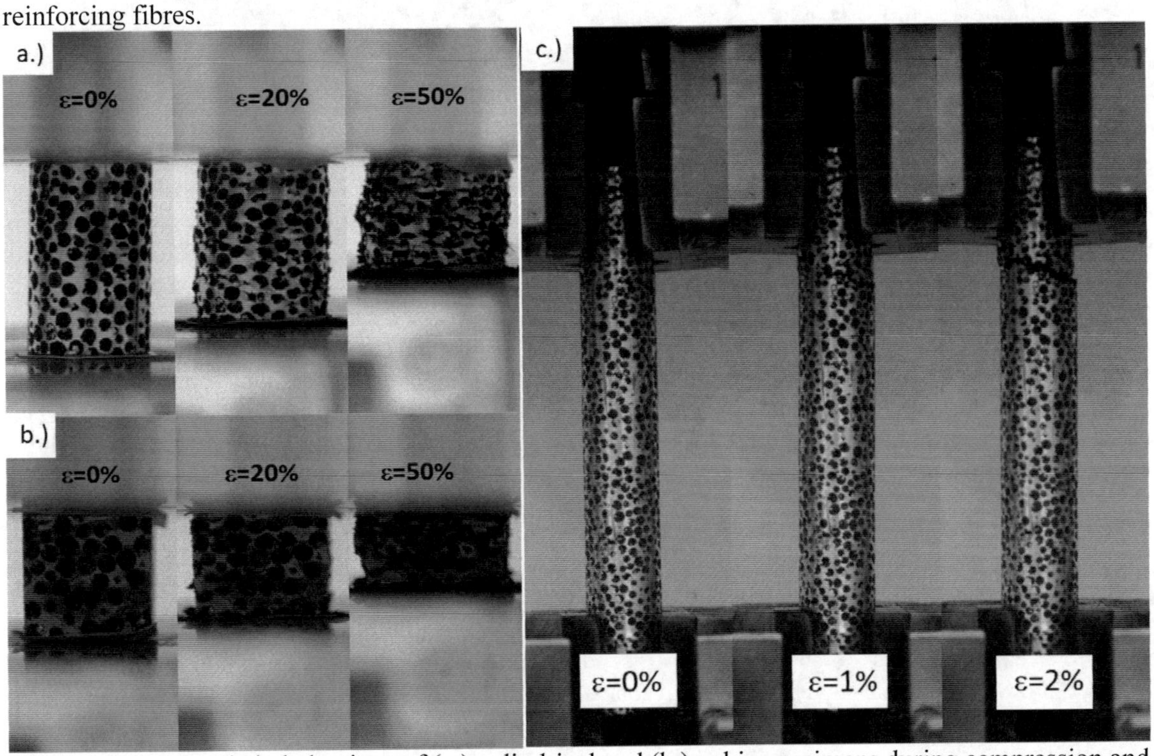

Figure 3. Characteristic behaviour of (a.) cylindrical and (b.) cubic specimens during compression and (c.) characteristic behaviour of tensile specimens (the indicated values show the engineering strain)

Compression tests were carried out with an MTS 810 universal material testing machine (MTS Systems, Eden Prairie, MN, USA) between two flat platens with a maximum load capacity of 250 kN. The speed of the compression was 4 mm·min^{-1} on the cylindrical specimens and 3 mm·min^{-1} on the cubical specimens, respectively. The tensile tests were carried out with a TIRATest 2300 universal testing machine (TIRA GmbH, Schalkau, Germany) with a maximum load capacity of 100 kN and a mechanical grip. The speed of the test was 4 mm·min^{-1}. Tensile tests were also carried out on individual CFBs with and without twisting with an Instron 5965 universal testing machine (Instron, Norwood, MA, USA) with a maximum load capacity of 5 kN and a cross-head speed of 3 mm·min^{-1}.

As the CFBs are not clearly visible in the cross-section of the foam, the tensile specimens were machined from the foam blocks without the exact information of where the CFBs were located. Therefore, the cross section of the CFBs was calculated. The maximum number of CFBs in a specimen was calculated based on the known spacing of the grid (2.4 mm). A CFB was placed at every other point on the grid, giving a spacing of 4.8 mm. Calculating with this 4.8 mm radius, a maximum of 13 CFBs can be placed in a Ø20 mm specimen, as shown in Fig 4 (a).

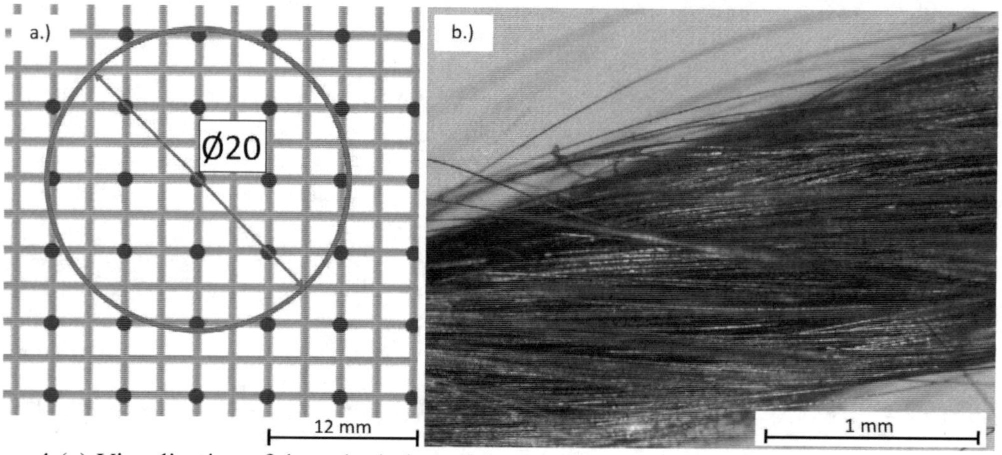

Figure 4.(a) Visualisation of the calculation of the maximum number of CFBs in a sample and (b) image of a twisted CFB used to measure diameter.

Images of the CFBs were taken using an Olympus SZX16 stereomicroscope (Olympus Corporation, Tokyo, Japan), and their diameters were measured using JMicroVision software. An image of a CFB is shown in Fig 4 (b). The average diameter of the CFBs was 1.28 ± 0.04 mm. Using the previous calculation and this value, the load-bearing cross section was determined as 16.68 mm^2 due to the high porosity of the metal matrix, making it not applicable against tensile loading.

The following mechanical properties were determined during the compression test: the plateau stress (σ_{pl}), which is the average stress between 20% and 30% engineering strain; the compressive offset stress ($\sigma_{p0.2}$), which is the stress at the engineering plastic strain of 0.2% and the absorbed energy up to 50% engineering strain (W_{50}) [22]. In addition, the density-specific version of these characteristics was investigated.

3. Results and discussion

The density of the samples was determined to be 1.57 ± 0.1 g·cm^{-3} (relative density around 0.58). The engineering stress– engineering strain curves of the tensile test are shown in Fig. 5a, and the curves of the CFBs are shown in Fig. 5b. As the twisted CFBs were able to withstand greater loads than those without twisting, it was concluded that it is advantageous to twist CBFs when using them as longitudinal reinforcement in syntactic metal foams.

The tensile strength of the specimens and the CFBs is shown in Table 1.

Table 1. The average tensile strength of the specimens

Sample	Twisted CFBs (MPa)	Non-twisted CFBs (MPa)
with the apparent cross-section	33 ± 2	22 ± 1
with the cross-section of the CFBs	424 ± 19	282 ± 17
CFBs	205 ± 31	130 ± 23

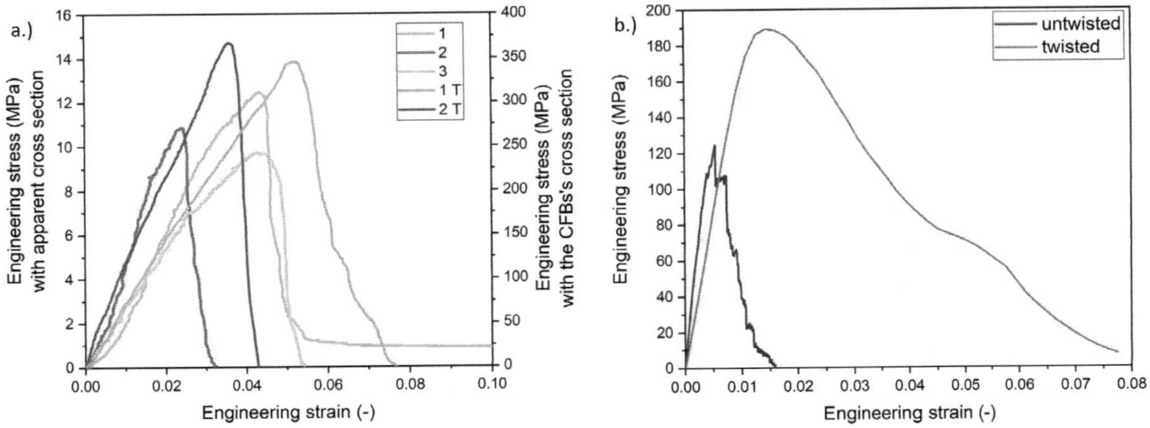

Figure 5. The engineering stress-strain curves of the tensile test calculated (a.) with the apparent cross-sectional area and (b.) with the cross-sectional area of the CFBs only (T: with twisted CFBs)

The values were calculated using the previously introduced method for the area of the CFBs cross-section, which uses a maximum number of CFBs in the cross-section and, therefore, gives a minimum value for tensile strength. Although it is a minimum value, the true tensile strength lies between the two calculated values.

The tensile strength of an individual carbon fibre is determined in the literature, ranging from 3000 to 7000 MPa [23], but these results could not be reproduced due to the fibres being in bundles. As shown in Fig. 5b, the fibres in the non-twisted bundle started to break one after the other, and the intact fibres took the load, while the fibres in the twisted bundle fractured at the same time and then the friction between the fibres was the force that took the load. The compressive engineering stress–enginering strain curves are shown in Fig. 6.

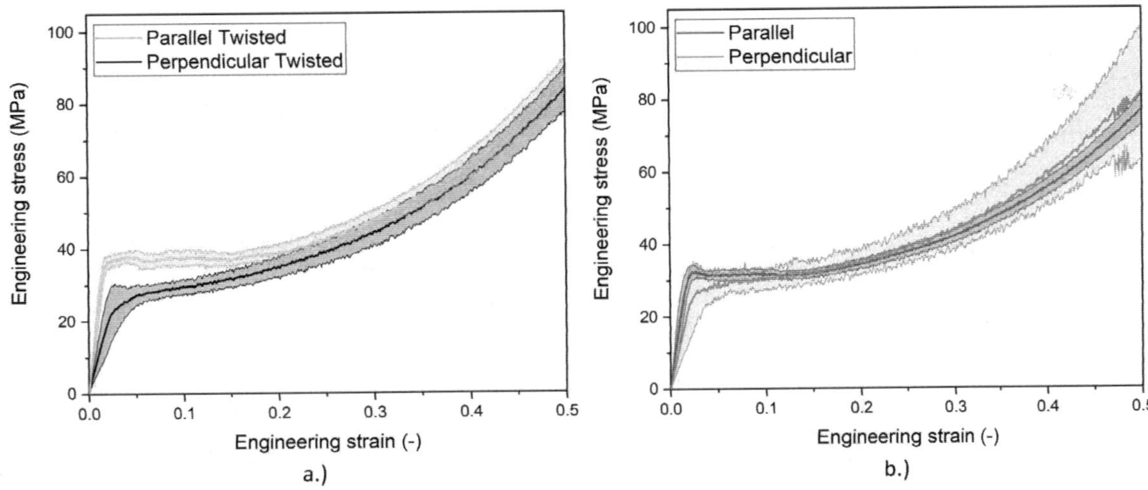

Figure 6. The measured and averaged engineering stress–engineering strain curves of the compression tests of the specimens with (a.) twisted and (b.) untwisted CFBs, with standard deviation

While the curves of the specimens compressed parallel to the orientation of the CFBs initially reach higher stress values, they fall below the curves of the specimens compressed perpendicular to the orientation of the CFBs in the later stages of compression. This means that the specimens compressed parallel to the CFB orientation have a higher compressive offset stress ($\sigma_{p0.2}$) but a lower plateau stress (σ_p) and absorb less energy in a unit of volume at an average up to 50% engineering strain (W_{50}). The mechanical properties are listed in Table 2. The difference between them is due to the fact that the

CFBs prevent deformation perpendicular to their orientation due to the bond and friction between them and the matrix material.

Table 2. The compressive mechanical properties

	Parallel	Parallel twisted	Perpendicular	Perpendicular twisted
Plateau stress (MPa)	37 ± 0.6	43 ± 1.3	39 ± 2,2	39 ± 1.6
Specific plateau stress ($MPa{\cdot}cm^3{\cdot}g^{-1}$)	23 ± 0.3	26 ± 0.7	25 ± 1.1	25 ± 0.8
Compressive offset stress (MPa)	32 ± 1.3	36 ± 0.9	28 ± 2.1	25 ± 2.9
Specific compressive offset stress ($MPa{\cdot}cm^3{\cdot}g^{-1}$)	20 ± 0.8	22 ± 0.5	18 ± 0.2	16 ± 1.8
Absorbed energy ($J{\cdot}cm^{-3}$)	21 ± 0.4	24 ± 0.7	22 ± 1.5	22 ± 0.9
Specific absorbed energy ($J{\cdot}g^{-1}$)	13 ± 0.2	15 ± 0,4	14 ± 0.8	14 ± 0.5

4. Conclusion

Metal matrix syntactic foams with unidirectional fibre reinforcement were produced successfully with a vacuum infiltration technique. The main cohesive force was the friction between the matrix and the fibres during mechanical testing. While the fibres in the non-twisted CFBs tended to break individually, their behaviour in the composite specimens was similar to that of the compressed bundles (in the specimens and individually) as they tended to break simultaneously. The fibre bundles resulted in advantageous properties when placed perpendicular to the direction of compression compared to when placed parallel to the direction of compression. During tensile tests, the fracture of the specimens occurred almost immediately, as expected, because the porosities of the syntactic foam matrix served as starting points for cracking. Meanwhile, during compression, the CFBs placed perpendicular to the direction of compression slowed down the lateral deformation of the specimens owing to the previously mentioned friction between the matrix and the fibres. The mechanisms behind the observed effects will be further investigated in the future.

5. Acknowledgements

This work was partially supported by the National Research, Development and Innovation Fund (grant number OTKA FK_21 138505). Project no. TKP-6-6/PALY-2021 has been implemented with the support provided by the Ministry of Culture and Innovation of Hungary from the National Research, Development and Innovation Fund, financed under the TKP2021-NVA funding scheme.

6. References

[1] Sivananthan S, Ravi K and Samuel S J 2020 Effect of SiC particles reinforcement on mechanical properties of aluminium 6061 alloy processed using stir casting route *Mater. Today Proc.* **21** 968–70

[2] Dhanashekar M, Loganathan P, Ayyanar S, Mohan S R and Sathish T 2019 Mechanical and wear behaviour of AA6061/SiC composites fabricated by powder metallurgy method *Mater. Today Proc.* **21** 335–41

[3] Ye T, Xu Y and Ren J 2019 Effects of SiC particle size on mechanical properties of SiC particle reinforced aluminum metal matrix composite *Mater. Sci. Eng. A* **753** 146–55

[4] Zaiemyekeh Z, Liaghat G H, Ahmadi H, Khan M K and Razmkhah O 2019 Effect of strain rate on deformation behavior of aluminum matrix composites with Al 2 O 3 nanoparticles *Mater. Sci. Eng. A* **753** 276–84

[5] Ucsnik S, Großalber A, Reiter J, Scheerer M and Mozdzen G 2019 Advances in continuous carbon fibre reinforced Aluminium matrix composites *8th Eur. Conf. Aeronaut. Sp. Sci. EUCASS_2019* **1** 2

[6] Khanna V, Kumar V and Bansal S A 2021 Aluminium-Carbon Fibre Metal Matrix Composites: A Review *IOP Conf. Ser. Mater. Sci. Eng.* **1033**

[7] Kumar N, Chittappa H C and Ezhil Vannan S 2018 Development of Aluminium-Nickel Coated Short Carbon Fiber Metal Matrix Composites *Mater. Today Proc.* **5** 11336–45

IOP Conf. Series: Materials Science and Engineering **1313** (2024) 012005 doi:10.1088/1757-899X/1313/1/012005

[8] Bouhamida B and Merzoug A 2021 Novel Analytical Model of Stress Concentration around Broken Fibers of Unidirectional Composite Plate *Period. Polytech. Mech. Eng.* **65** 120–8
[9] Bedmar J, Torres B and Rams J 2022 Manufacturing of Aluminum Matrix Composites Reinforced with Carbon Fiber Fabrics by High Pressure Die Casting *Materials (Basel).* **15**
[10] Yang Q, Liu J, Li S, Wang F and Wu T 2014 Fabrication and mechanical properties of Cu-coatedwoven carbon fibers reinforced aluminum alloy composite *Mater. Des.* **57** 442–8
[11] Miyoshi T, Itoh M, Akiyama S and Kitahara A 2000 ALPORAS aluminum foam: Production process, properties, and applications *Adv. Eng. Mater.* **2** 179–83
[12] Marx J and Rabiei A 2017 Overview of Composite Metal Foams and Their Properties and Performance *Adv. Eng. Mater.* **19**
[13] Mirzarahimi M K, Khodarahmi H, Ziashamami M and Hosseini R 2023 Numerical and Experimental Study of the Use of Mineral Pumice in the Core of the Sandwich Panel to Absorb the Shock Wave *Period. Polytech. Mech. Eng.* **67** 19–29
[14] Jang W Y and Kyriakides S 2009 On the crushing of aluminum open-cell foams: Part I. Experiments *Int. J. Solids Struct.* **46** 617–34
[15] San Marchi C and Mortensen A 2001 Deformation of open-cell aluminum foam *Acta Mater.* **49** 3959–69
[16] Hangai Y, Ando M, Ohashi M, Amagai K, Suzuki R, Matsubara M and Yoshikawa N 2020 Compressive properties of two-layered aluminum foams with closed-cell and open-cell structures *Mater. Today Commun.* **24** 101249
[17] Santa Maria J A, Schultz B F, Ferguson J B and Rohatgi P K 2013 Al-Al2O3 syntactic foams - Part I: Effect of matrix strength and hollow sphere size on the quasi-static properties of Al-A206/Al2O3 syntactic foams *Mater. Sci. Eng. A* **582** 415–22
[18] Kádár C, Kubelka P and Szlancsik A 2023 On the compressive properties of aluminum and magnesium syntactic foams: Experiment and simulation *Mater. Today Commun.* **35**
[19] Jenei P, Kádár C, Szabó Á, Hung S M, Kuo C J, Choe H and Gubicza J 2022 Mechanical behavior of freeze-cast Ti foams with varied porosity *Mater. Sci. Eng. A* **855**
[20] ISO 6892-1:2019 Metallic materials - Tensile testing - Part 1: Method of test at room temperature
[21] Standardization I O for 2011 Mechanical testing of metals – Ductility testing – Compression test for porous and cellular metals *Int. Organ. Stand.* **2011** 1–7
[22] Szovák B, Károly D, Kemény A, Miklós D and Kádár C 2019 Materials Today: Proceedings *Mater. Today Proc.* **37** 0–5
[23] Mirdehghan S A 2021 Fibrous polymeric composites *Eng. Polym. Fibrous Mater.* 1–58

IOP Conf. Series: Materials Science and Engineering **1313** (2024) 012006
doi:10.1088/1757-899X/1313/1/012006

The effect of forming conditions and treatment process applied to the material on springback

J E Dessie and Z Lukács

Institute of Materials Science and Technology, Faculty of Mechanical Engineering and Informatics, University of Miskolc, H-3515 Miskolc, Egyetemváros, Hungary

E-mail: metjemal@uni-miskolc.hu

Abstract. Heat treatment is often used to optimize the balance between the strength and formability of a material. However, aging during treatment affects the ability to return to its original shape after deformation. In this study, the effects of surface treatments during the painting of mild steel and aging time in the W-temper heat treatment on springback were investigated. Consequently, because of artificial aging during surface treatment, the material precipitates certain phases and enhances the mechanical properties of the material, which leads to the pre-painted mild steel being more sensitive to springback. On the other hand, it is also noted that springback is highly dependent on the transfer time between the W-temper heat treatment of high strength aluminum alloy and forming operations, and the change in shape increases with time. In summary, the springback phenomenon is a result of the interaction between the material properties, treatment procedures, and forming circumstances.

1. Introduction

The phenomenon known as springback happens to materials when they are deformed and subsequently released, especially for metals. It describes a material's propensity, either fully or partially, to revert to its initial shape following deformation [1,2]. Springback can be affected by the deformation technique applied to the material, such as stamping, bending, or drawing. The amount of springback is dependent on forming conditions, die design and tooling. Overbending stamped parts is the most widely used technique to prevent this impact. Such flaws need to be recognized, prevented, or removed. Their precise causes and connections to other influencing factors also need to be investigated and comprehended. It is believed that the springback phenomenon is largely produced by the distribution and amplitude of stresses generated during sheet thickness creation [1,3,4]. For springback research to be accurate, a material model that can precisely describe the intricate material behavior during unloading is required [5, 6]. Apart from the sheet's material properties, the friction coefficient of the contact surfaces and the blank holding force also frequently affect springback. There are further effects from the sheet thickness and the tool profile radii [7]. As a result, a detailed examination of the different ways in which these factors impact springback is essential and necessitates a high degree of proficiency with a wide range of experimental design matrix variables.

Pre-painted products are still in high demand, and because pre-painted sheets are more affordable, more environmentally friendly, and result in better-looking surfaces, more industries have started employing them [8]. Furthermore, manufacturers do not require an in-plant painting operation because they may readily obtain pre-painted sheets from external providers [9].

To increase high strength aluminum alloy sheet's formability, the W-temper forming ideal method, which is entails heating it to a specific temperature and then quickly cooling it. Nevertheless, during W-temper (WT) production, the sheet is created at ambient temperature, and springback remains a difficult technical problem. The link between different W-temper process parameters, including temperature, cooling rate, and holding time, has been the subject of numerous investigations. Using both mathematical modeling and testing, Gronostajski et al. [10] found the ideal WT condition for producing the B-pillar from 7075 aluminum alloy. The most significant issues that come up during the cold forming of solution-heat-treated sheets and the choice of ideal cooling and temperature settings were the main topics of this study. Using well-modeled constitutive equations, Choi et al. [11] conducted a study on the mechanical properties of a WT high-strength aluminum alloy and used the results in finite element (FE) simulations. The evaluation of springback in WT forming was conducted using the U-draw bending test. In the numerical modeling of springback for possible application in the cold-forming process, anisotropic plasticity under loading path variations was highlighted in terms of yield function and hardening law. Lee et al. [12] looked at the mechanical behavior of WT sheets made of 7075 aluminum alloy and assessed whether it would be possible to use the WT forming technique to produce exact auto parts. To assess the fundamental mechanical properties and formability of plastic and elastic materials, mechanical experiments such as the Nakajima test and uniaxial and balanced biaxial tension tests were performed. Using three-point bending experiments, Wang et al. [13] thoroughly investigate the effects of heat treatment factors on the stamping deformation and springback of 6061 aluminum alloy sheets. Additionally, the mechanism of the sheet springback was investigated by merging it with FE numerical modeling. Overall, the literature on this subject has shown that springback during metal forming operations can be effectively reduced by the WT forming process, but more investigation is required to fully comprehend the connection between springback and the parameters of the WT process.

In this study, the effect of surface treatment during the painting of DC01 steel before stamping and the effect of transfer time in the W-Temper forming process of AA6082 aluminum alloy on springback were investigated. Experimental and numerical studies have been conducted. The numerical analysis was conducted using the commercial AutoForm code.

2. Material Characterization

2.1. Material characteristics of unpainted and pre-painted steel

The mechanical properties of the unpainted (DC01) and pre-painted (DX51D) mild steel were assessed using a uniaxial tensile test. Table 1 provides a summary of the formability and mechanical performance of painted and unpainted sheets. Since the unpainted sheet lacked a defined yield strength in these circumstances, the offset yield strength was determined. The test determined the upper and lower yield points, and it reduced by 9%. However, the pre-painted sheet only specifies the tensile strength and not the yield strength. The tensile diagram of this kind shows that the sheet aged after rolling. Comparatively speaking, the pre-painted steel's upper yield strength at the end of the elastic portion was much higher (nearly 1.5 times) than the unpainted steel's. A ratio of 0.6 for the unpainted and 0.8 for the pre-painted is found when comparing the yield strength and tensile strength ratios; these values are nearly equal.

Table 1. Mechanical and formability behavior of unpainted and painted sheets

Material type	Elastic Modulus, E_0, GPa	ReL, Mpa.	ReH, Mpa.	Rp0,2, Mpa.	Rm, Mpa.	strain hardening exponent (n)	Total elongation, A_g, (%)
DC01	206	-	-	192	322	0.194	24.5
DX51D	206	264	290		360	0.133	20.3

2.2. Tension and unloading tests of W-temper aluminium alloy

The commercially produced peak-aged (T6 tempered) 6082 aluminum alloy sheets fail early during the forming process at room temperature because they are sometimes very hard to form while being deformation. WT forming's primary benefit is that it increases formability at room temperature, enabling conventional cold stamping. Because WT sheets are more formable, it has been thought that they might be produced at room temperature. On the other hand, springback after cold stamping may be a significant issue. For this reason, it's critical to examine the Bauschinger effect under cyclic loading and the apparent elastic modulus's deterioration, to precisely forecast the springback in the numerical simulation. Uniaxial tension-unloading tests of W-tempered AA 6082 were carried out to achieve this. To ensure that a homogenous temperature distribution and all the alloying components had fully dissolved into the aluminum matrix, the 6082-T6 aluminum alloy was heated to 525°C for 30 minutes [14]. It was then cooled to room temperature using water. This heat treatment is being researched since SHT is known to improve ductility. From the test, the flow stress and apparent elastic modulus deterioration were examined at about 15 and 120-minute aging times (the transfer period from quick quenching to formation). As seen in Figures 1 and 2, it is evident that a notable variation in the flow stress and a decrease in the elastic modulus were noted.

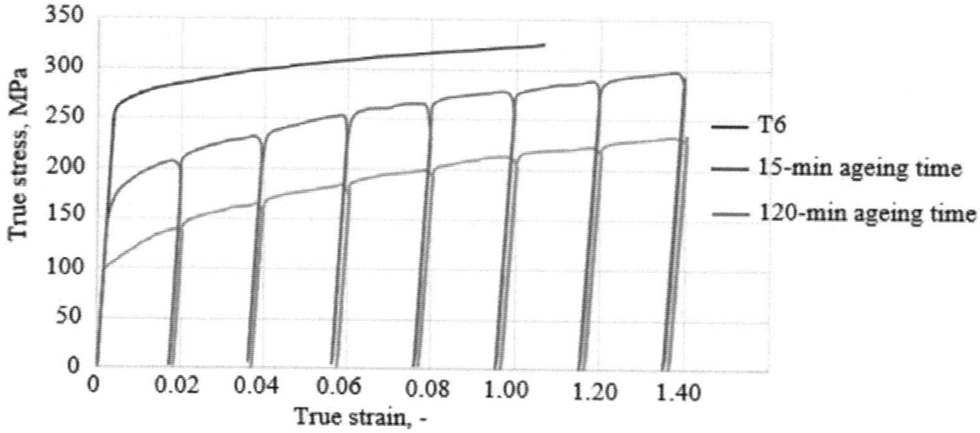

Figure 1. Tension-unloading test

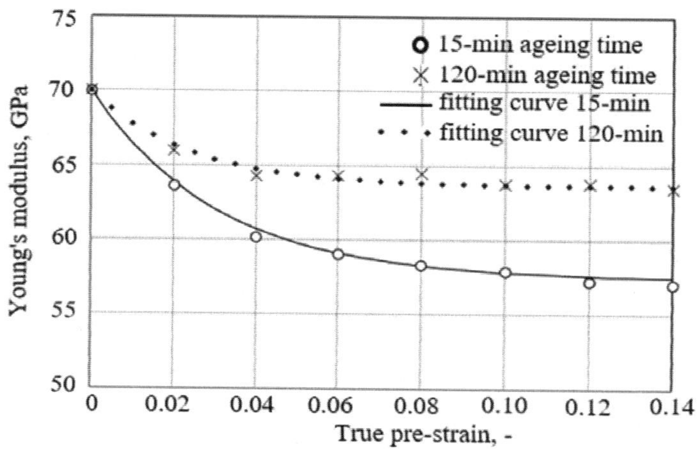

Figure 2. Reduction of elasticity modulus as a function of pre-strain

3. Finite Element Modelling

Using the AutoForm commercial code, U-draw bending of 250x20x0.45 mm for unpainted and pre-painted DC01 steel as well as 250x20x1 mm for WT AA 6082 aluminum alloy was done numerically. A fine triangular shell element was employed because of the numerical simulation of an extremely thin sheet. Figure 3 displays the model used in AutoForm for the numerical simulation.

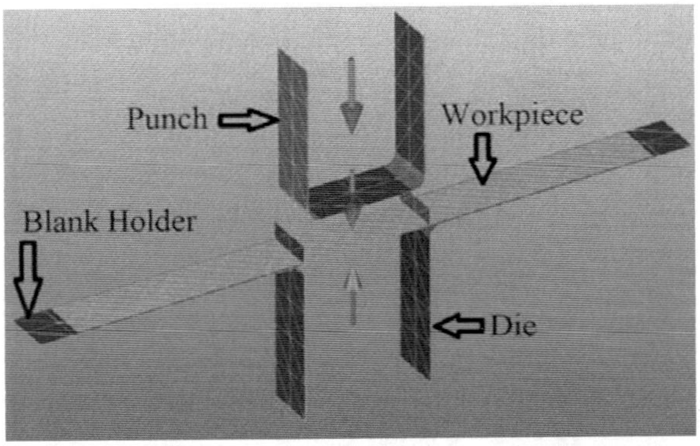

Figure 3. Tension-unloading test

A novel approach has been devised and implemented in the AutoForm commercial code to represent the material's kinematic hardening behavior as Eq. (8) [15].

$$E_l = E_0(1 - \gamma(1 - e^{-\chi p})) \quad (8)$$

where E_l is the tangent modulus which ordinarily falls off exponentially as comparable plastic strain, E_0 is the tangent modulus at zero plastic strain, χ is saturation constant, γ is Young's reduction factor.

This study only considers the kinematic hardening parameters that are already pre-defined in the material card of AutoForm for unpainted and pre-painted steel materials, as shown in Table 2.

Table 2. Kinematic hardening parameters

γ	χ
0.24	40

The γ and χ values in the numerical simulation of W-temper formation can be discovered by fitting curves to these observed points with the MATLAB program's curve fitting module [16], as seen in figure 2. It can be demonstrated that the average chord modulus value falls with increasing deformation. Consequently, Table 3 displays the kinematic hardening parameters that were ascertained from the fitting curve for the 15 and 120-minute ageing durations WT formation.

The constant process parameters that were considered for the numerical analysis of all causes are listed in Table 4.

Table 3. Kinematic hardening parameters.

Aging time	γ	χ
15-minute	0.18	33.3
120-minute	0.091	43.1

Table 4. Design variables and their level for the simulation.

Parameters	Level
The radius of die and punch profile	5mm
Blank holding force (BHF)	3, 5, 7kN
μ	0.08

4. U-draw bending test

In the numerical investigation, lubricated conditions were used for U-bending test trials. The experimental configuration of the U-bending dies apparatus, mounted on a 20kN hydraulic press machine, is depicted in Figure 4. The die clearance was double the thickness of the sheet on both sides. Table 5 displays the constant variables and their respective levels for the experiment.

Figure 4. U-bending test apparatus.

Table 5. Design variables and their level for the experiment.

Parameters	Level
Radius of die and punch profile	5mm
BHF	3, 5, 7kN
Lubrication condition	oil lubrication

5. Results and discussion

The amount of springback of each test was measured using NUMISHEET '93 benchmark standard as it is shown in Figure 5.

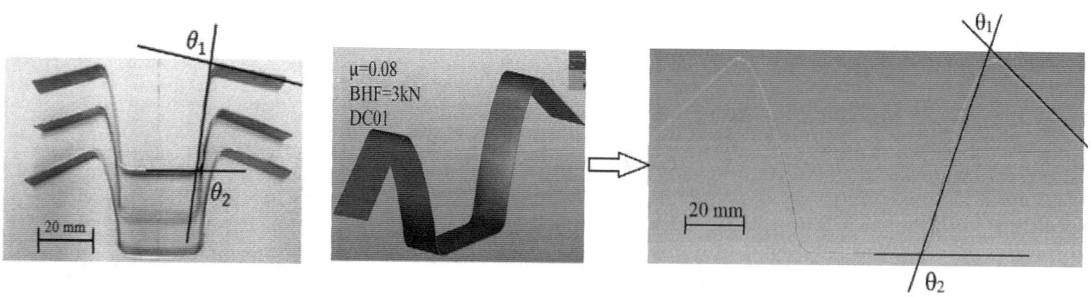

Figure 5. NUMISHEET '93 benchmark standard for springback prediction [17]

5.1. Effect of pre-painting before stamping on springback

For the unpainted and pre-painted steel, Figure 6 displays the springback amounts as a function of the BHF in both the experimental and numerical simulations. All the computational and experimental tests of both materials show inverse connections between the deviation of angles from the origin and the BHF. With increased BHF, all springback angles are reduced due to the more even distribution of strains across the thickness of the sheet. It can be observed that, particularly at higher BHF levels, the springback angles derived from the numerical analysis for unpainted steel are in moderate agreement with the experimental results. Because the numerical simulation did not account for the impact of paint materials on the coefficient of friction, the springback parameters for pre-painted steel are somewhat different from those observed in the experiment. As a result, further adjustments to the numerical simulation are needed to account for the impact of paint material on the coefficient of friction.

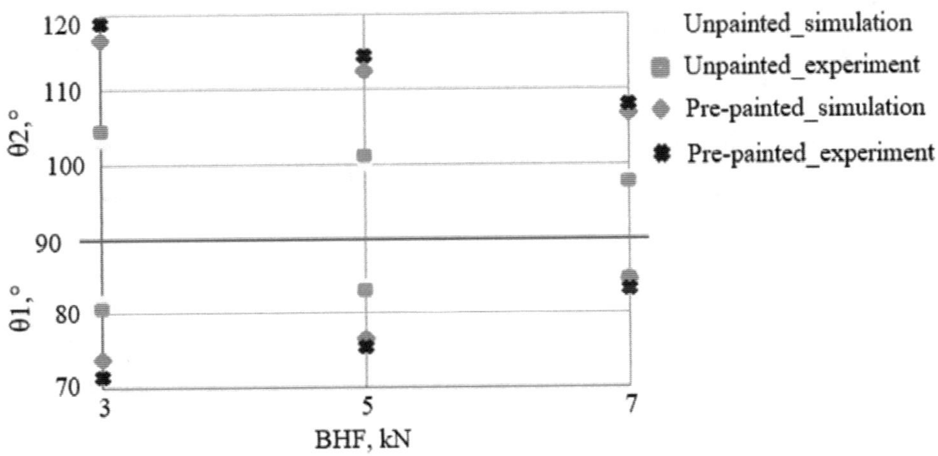

Figure 6. Effect pre-paint on springback as a function of BHF

5.2. Effect of Transfer time in W-temper forming on springback

The amount of time that 6082 WT aluminum alloy transfers to the forming process can have an impact on the alloy's springback. A higher springback angle may arise from a longer transfer period between the W-temper heat treatment and the stamping process since this gives the material more time to take on its final shape as it ages. The relationship between the angle deviation θ_1 and the transmission time is inverse. On the other hand, Figure 7 illustrates a direct relationship for θ_2. Increasing the ageing period had no further influence on springback angle since the material may naturally age to some amount, which may improve its strength while lowering its ductility. As a result, it's critical to carefully optimize the aging duration for the particular application and material being used, while also weighing other considerations like manufacturing efficiency and cost.

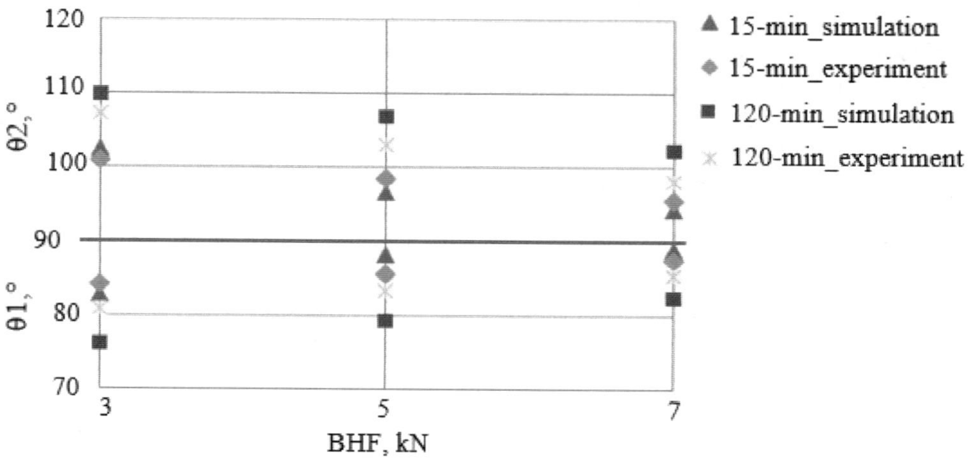

Figure 7. Effect of edging time on springback angle as a function of BHF

The transfer time could be impacted by the BHF. Figure 8 illustrates how springback diminishes with increasing blank holding power. This is because the blank holding force helps to evenly distribute the deformation throughout the sheet metal during the forming process. As a result, there are fewer residual stresses and a more even distribution of stresses, which decreases the tendency of the material to return to its original shape after forming. It is important to keep in mind that exceeding a certain blank holding force point can result in additional problems such as excessive material thinning or tearing. Therefore, it is necessary to determine the optimal blank holding force based on the specific properties of the material and the forming conditions.

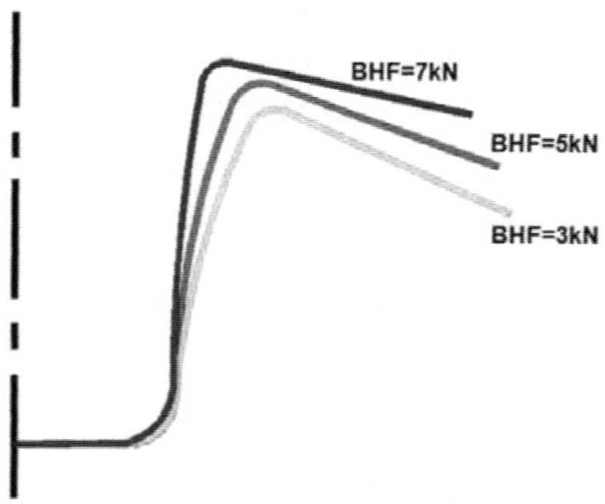

Figure 8. Effect of BHF at 15-minute edging time

6. Conclusions

The effect of a heat treatment process on springback depends on the specific treatment applied and the material used. Surface coatings and treatments have an impact on the whole deformation process, springback behavior, and material properties. The pre-painted steel sheet is more affected by springback parameters in the U-draw bending test than the unpainted steel sheet. The W-Temper forming of high-strength aluminium alloys is a highly effective heat treatment process that can significantly improve the mechanical properties and performance of these materials. The interplay of material characteristics, treatment processes, and formation conditions leads to precisely predicting the springback phenomena. To achieve desirable forms and dimensions in the product, engineers and manufacturers must carefully evaluate these elements and use appropriate ways to limit or regulate springback.

References

[1] Brabie G, Chirita B, Nanu N and Ciubotariu V 2009 Analysis of the springback and residual stresses generated by cold plastic forming in drawn round parts made from steel sheets *Metalurgia International* **14(12)** 21-27

[2] Samuel M 2000 Experimental and numerical prediction of springback and side wall curl in U-bendings of anisotropic sheet metals *J. Mater. Process. Technol.* **105(3)** 382-393

[3] Behrens B. A, Yun J. W, Schäffner C, Kroening M, Altpeter I, and Kopp M 2005 Influences on the part quality in conventional deep drawing processes *International conference on steels in cars and trucks* Wiesbaden Germany 144-151

[4] Asnafi N 2001 On springback of double-curved autobody panels *Int. J. Mech. Sci.* **43(1)** 5-37

[5] Aday A J 2019 Analysis of Springback Behavior in Steel and Aluminum Sheets Using FEM *In Annales de Chimie: Science des Materiaux* **43(2)** 95-98

[6] Yang X, Choi C, Sever N.K and Altan, T 2016 Prediction of springback in air-bending of Advanced High Strength steel (DP780) considering Young' s modulus variation and with a piecewise hardening function *Int. J. Mech. Sci.* **105** 266-272

[7] Dessie J and Lukacs Z 2023 Determination of influential springback parameters in U-bending test Pollack Periodica **18(2)** 17-22

[8] Nguyen D, Kim Y and Jung D 2012 Finite element method study to predict spring-back in roll-bending of pre-coated material and select bending parameters *Int. J. Precis. Eng. Manuf.* **13(8)** 1425-1432

[9] Ciccarelli D, Forcellese A, Greco L, Mancia T, Pieralisi M, Simoncini M and Vita A 2021 Deformation behavior of pre-painted steel sheets *Procedia CIRP* **99** 266-27

[10] Gronostajski Z, Jaśkiewicz K, Kaczyński P, Skwarski M, Polak S, Krawczyk J, Chorzępa W and Trzpis P 2022 W-temper forming of B-pillar from 7075 aluminum alloy *CIRP Annals* **71(1)** 221-224

[11] Choi Y, Lee J, Panicker S.S, Jin, H.K, Panda S.K and Lee M.G 2020 Mechanical properties, springback, and formability of W-temper and peak aged 7075 aluminum alloy sheets: Experiments and modelling *International Journal of Mechanical Sciences* **170** 105344

[12] Lee J, Bong H.J, Kim D, Lee Y.S, Choi Y and Lee M.G 2019 Application of combined W-temper and cold forming technology to high-strength aluminum alloy automotive parts *JOM* **71** 4393-4404

[13] Wang G, Zhu G, Kou L, Li T, Liu Z, Shang X, Jiang X and Zhu X 2019 Effect of heat treatment conditions on stamping deformation and springback of 6061 aluminum alloy sheets *Materials Research Express* **7(1)** 016512

[14] Dessie J and Lukacs Z 2023 Experimental investigation of the effect of W-temper forming process parameters on springback *In IOP Conference Series: Materials Science and Engineering*, **1284(1)** 012066

[15] Kubli W, Krasovskyy A, and Sester M 2008 Modeling of reverse loading effects including workhardening stagnation and early re-plastification *International Journal of Material Forming* **1(1)** 145–148

[16] Hegedűs G 2017 Application of knowledge-based design in computer aided product development *In Vehicle and Automotive Engineering: Proceedings of the JK2016* Miskolc Hungary 109-114

[17] Makinouchi A 1993 NUMISHEET'93 *Proceedings of the 2nd International Conference Numerical Simulation of 3-D Sheet Metal Forming Processes; Verification of Simulation with Experiment* Isehara Japan

IOP Conf. Series: Materials Science and Engineering **1313** (2024) 012007 doi:10.1088/1757-899X/1313/1/012007

Impregnation problems of aluminum castings

M Réger, R Horváth*, ER Fábian and T Réti

Bánki Donat Faculty of Mechanical and Safety Engineering, Óbuda University
Népszinház u 8, Budapest, H-1081 Hungary
*horvath.richard@uni-obuda.hu

Abstract. Vacuum impregnation is an effective solution to the porosity problem of aluminum die castings. Experience has shown that quality die-cast aluminum alloy products manufactured with gas tightness requirements can show leakage even after impregnation operation. The study seeks to answer the question of the geometry of the cavity system through which leakage occurs and why the cavity system is not saturated with liquid resin during impregnation. It is assumed that insufficient impregnation occurs if the near-surface parts of the cavity system are heavily deformed locally during the machining of the casting.

1. Introduction

Impregnation with resin is an effective method of sealing leak paths in aluminum alloy castings manufactured with gas tightness requirements. During impregnation, the impregnating material used fills and cures, sealing the internal interconnected porosity cavity system without altering the size or functional properties of the casting [1].

One of the earliest descriptions of the impregnation of porous castings can be found in Vernon C. F. Holm's research report [2]. He was experimenting with a resin-based sealant to seal porous metal casting cavities. Because of its military importance, impregnation has been subject to military standards since the 1950s.

Today, vacuum impregnation is an effective solution to the porosity problem of aluminum castings [3]. In this process, after the parts have been vacuum treated, in order to ensure a permanent seal, the porosity cavities are filled with a thermosetting polymer that is resistant to the expected influences. First, the castings to be impregnated are placed in a cage so that all surfaces of the product can come into contact with the surrounding medium (gas, liquid), and next the surfaces are cleaned by washing. The impregnation itself is carried out in a special autoclave, the lower part of which is pre-loaded with the liquid impregnating resin, but the parts are first placed in the upper part of the autoclave. During the vacuuming of the system (vacuum level below 1 mbar, duration of about 10 minutes), moisture and most of the air are also removed from the internal cavities of the castings. While vacuuming is still in progress, the cage holding the parts is dipped into the impregnating resin and the autoclave is then opened. This part of the operation takes about 5 to 6 minutes, during which the atmospheric pressure acting on the liquid surface of the resin effectively promotes the penetration of the impregnant into the cavities. This is followed by cold and hot water wash. The latter one results in curing of the resin in the cavities (around 90-95° C). Intensive washing is necessary because vacuum impregnation is carried out when the parts are already fully finished, i.e. no impregnant residues should remain on the surfaces. Figure 1 shows the variation of pressure and temperature during the vacuum impregnation process.

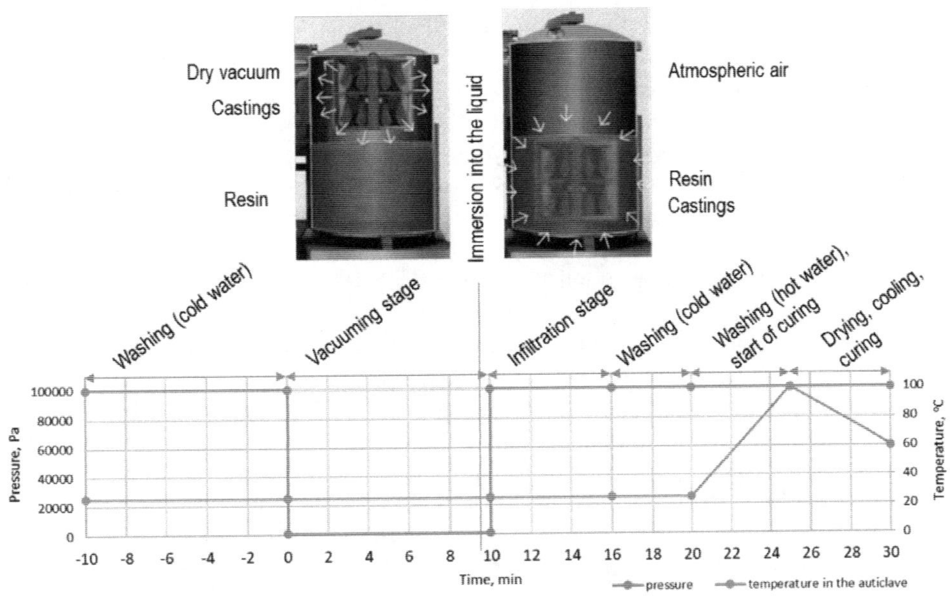

Figure 1 Steps of the vacuum impregnation process

Using vacuum impregnation technology, microporosities, leakage hairline cracks and other discontinuities of the machined aluminum castings that are open towards the surface can usually be reliably sealed. The resin in the discontinuities is transformed by heat treatment into a flexible solid material that is resistant to pressure, vibration and chemical influences.

A commonly used sealant for impregnating aluminum castings is e.g. Ultraseal MX2, a high-performance sealant approved by most automotive manufacturers and numerous multinational companies [4]. A representative of recyclable sealants is e.g. Rexeal 100, which offers substantial environmental advantages compared to conventional single-use non-recyclable impregnants (process stability, catalyst consumption, washing properties, thermal resistance, environmentally friendly application, etc.). Both impregnants comply with the internationally accepted US military specification for impregnation of castings and powder metallurgy products [5]. Basically this study seeks to answer the question of the geometry of the cavity system through which leakage occurs and why the cavity system is not saturated with liquid resin during impregnation.

2. Possible causes of insufficient impregnation

The impregnation process starts with vacuuming, partially to remove moisture and air from the open cavities and also to reduce pressure. In an ideal case, the pressure of the gas remaining in the cavity at the end of the vacuuming stage is close to the pressure level in the autoclave. Depending on the geometry, permeability and volume of the cavity system, equalization of pressures may not necessarily occur and pressure higher than the external vacuum level may remain in the internal cavities. The success of the impregnating resin saturation process is dependent on the achievement of an adequate vacuum level in this step, since the driving force for filling the cavity with impregnating resin is, in addition to the capillary effect, the difference between the external atmospheric pressure and the pressure created in the cavity by vacuum. In addition to the pressure difference, the saturation of the cavity system with resin depends essentially on the permeability of the cavity system. The slowing down of the flow rate of resin, i.e. the increase in flow resistance, is most likely caused by bottlenecks in the cavity system, since it is inversely proportional to the fourth power of the radius of the cross-section. From an impregnation point of view, the most unfavourable case is when the bottleneck is formed at or near the surface. Saturation will continue until the external and internal pressures are equalized or the time available in the technology has elapsed and the casting is removed from the liquid resin.

There is little available, reliable test information in the literature on the impregnation efficiency of complex geometry cavity systems. According to laboratory impregnation experiments on sintered metal alloys with controlled pore structure, cavities smaller than 500 μm are most likely saturated with resin, and pores smaller than 100 μm are always saturated with resin [3]. Based on the experience of the preliminary studies and the relevant calculations that formed the basis of the present work, it is likely that for aluminum alloy castings produced with high quality, gas tightness requirements, the impregnation of cavities containing small cross-sectional sections rather than large ones can be a problem [5].

Test experience has shown that the leakage of impregnated parts always occurs via a leakage path between machined surfaces, i.e. the internal cavity system becomes a through cavity system as a result of the machining process - in line with theories described in the literature [7]. A further observation is that leakage is often observed in the internal bores, on the surfaces formed by the thread rolling. The question is why the otherwise very effective impregnation technology sometimes fails to seal the leakage paths between the machined surfaces.

One possible explanation is that, for a given casting, the impregnation process is locally hindered by some random effect (e.g. contamination is introduced on the leaking surface that prevents the cavity from vacuuming or saturating). We believe that a system-level impregnation problem can occur in cavities with narrowed ends running out to the surface, e.g., due to strong plastic cold forming. The reduction of the cross-section at both ends of the through cavity system can also delay the vacuuming and saturation of the cavity system.

Each machining operation (turning, grinding) produces some plastic cold deformation in the near-surface material part, which can extend to a few hundredths or even a few tenths of a mm from the surface [8][9]. Thread rolling is a plastic cold forming operation that produces deformed condition on the surface of the material being machined to a depth of the order of the thread depth (several tenths of a mm). Although the plastic deformation in each of these operations is inhomogeneous in nature and decreases in magnitude away from the surface, it is certain that the surface run-out ends of the leakage cavity system do undergo some degree of plastic deformation. This can also result in a reduction in the cross-section of the leaking cavity in the vicinity of the surface. Results from preliminary calculations also show that a short (50-150 μm) cavity section with a diameter of a few microns can drastically reduce both vacuum and saturation efficiency.

3. Determining the characteristics of a leaking cavity system

The impregnability, and thus the leakage phenomenon, is essentially determined by the characteristics of the interconnected cavity system that forms the leakage path. This cavity system is an complex three dimensional shape, consisting of a chain of cavities of variable size and volume, and is permeable to gases along its entire length. Due to the spatial complexity and, occasionally, the order of magnitude size differences between the individual elements, it is generally not possible to map the interconnected elements of the system, i.e. to determine the exact leakage path. However, the distribution and density of discontinuities, which can be identified by destructive and non-destructive testing methods, can help to identify the probable location of the leakage path. Porosity can be regarded as a set of spherical voids characteristic of gas inclusions, and cavities of irregular geometry formed as a result of shrinkage. In quality castings, for example gastight castings, the equivalent diameter of the individual cavities is between a few μm and a few tenths of a mm based on micro CT [10][11][12][13][14] and metallographic[15][16][16][18] tests. In a volume of 1 mm3, sometimes up to 50–70 cavities are detected.

Detecting the real connected network in the cavity system is problematic with both destructive and non-destructive methods, as it is a complex spatial network of capillary-like channels with a small cross-section and dimensions below the resolution of CT[13][15][18].

An attempt has been made to determine the geometric characteristics of the leakage cavity system for a compressor connection stub (Al = 88 wt%; Si = 12 wt%) which has been proven to leak. The assembled compressor was screened for helium leakage, and the leak was identified through the threaded bore of the stub on the outer atmosphere side. After the leaky stub section was machined out, X-ray (YXLON FF35) images were taken from multiple directions to determine the approximate location of the leak cavity system. In Figure 2a, the band near the right side bore where the cavities cluster is clearly identified. The image perpendicular to this plane (Figure 2.b) also shows that the cavities are mostly located along the centerline of the investigated part (marked with a circle). To reveal the critical cross-section identified, the sample was notched in the appropriate directions on its lateral surfaces and then split. The resulting fracture surface contains a large number of cavities, which allow at least partial identification of the geometry of the leaking cavity system. Figure 3a shows a 3-dimensional optical micrograph of the fracture surface, in which the chain of larger cavities in the central part of the specimen is clearly visible. More detailed optical (Olympus DSX-1000) and scanning electron microscopic (JEOL JSM 5310 LA) investigations also suggest the location of the branches of the cavity system running out to the surface (Figure 3b), but of course the possibility of other traces cannot be excluded, mainly due to the spatial extent of the cavity system.

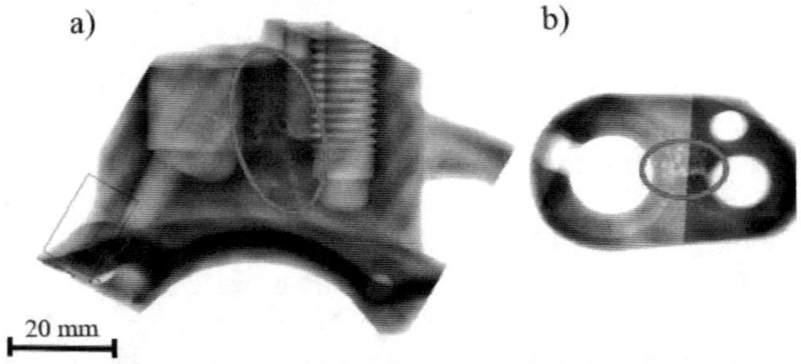

Figure 2 X-ray images of the leaking part

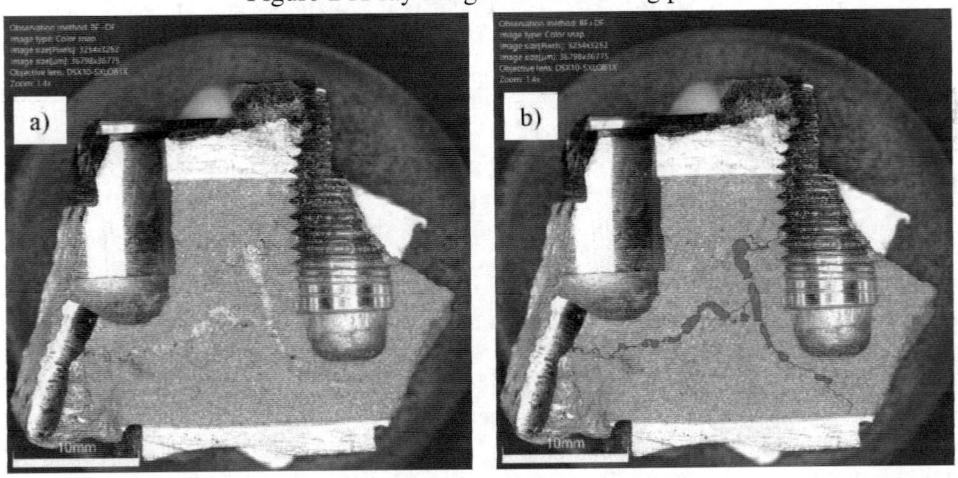

Figure 3 The artificial fracture surface of the leaking part (a) the original picture, b) theoretical leakage path)

Figure 4 shows the location of the cavities identified on the surface and the channels connecting them. The size and approximate volume of these formations can well be estimated by optical microscopy. On this, not entirely planar section, two types of leakage paths are plausible, using the lettering in Figure 4 along the ABC or ABD traces. The total cavity volume identified along the trails is 10.6 mm^3 between points ABC and 12.7 mm^3 along the ABD trail. The average volume of cavities along the trail is 1.4 mm^3, with the largest cavity estimated to be over 6 mm^3. The average diameter of the channels connecting the cavities is around 180 μm. We estimate that the diameter of the channel exceeds 20 μm everywhere.

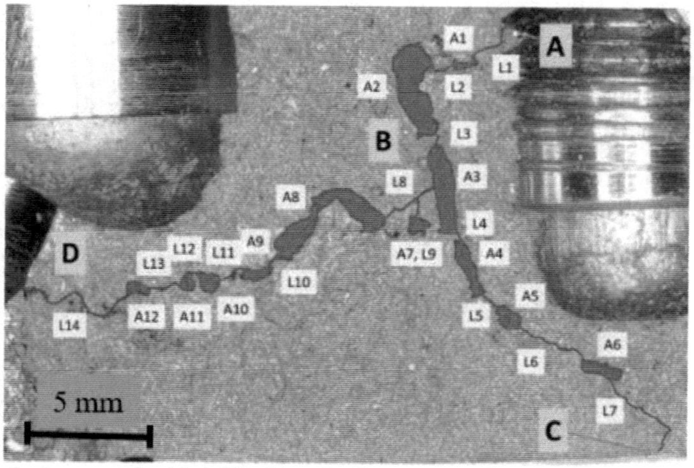

Figure 4 Possible leakage paths identified on the fracture surface (*A* is surface of the elementary capillary and the *l* the length of elementary capillary section)

As discussed in the previous chapter, the geometry of the connection of the cavity system to the surface is of paramount importance for the impregnation process. The helium leaking inspection identified leakage in the threaded hole shown in Figures 3 and 4. To identify the possible exit points within the bore, scanning electron microscopy investigation was performed on the threaded section. In the vicinity of the threaded profile shown in Figure 5.a, several shrinkage cavities are visible on the surface and dendritic patterns were observed on larger surface areas (Figure 5.c and 5.d), We hypothesized that this part of the threaded hole could (also) be the exit point of the leak channel. Strong deformation traces corresponding to the deformation can be identified on the surface of the thread profile (Figure 5.b), of which the deformation trace marked with an arrow can be identified as the surface section of the leakage channel. Figure 5.b shows the detail of 5.a at higher magnification. Because of the strong deformation, the internal "diameter" of the channel-like formation indicated by the arrow is between 1-1.3 μm, with a length of about 60 μm as shown in the image. Note that the height of the thread-tooth, i.e. the thickness of the actually deformed region, is more than half a mm. Due to the mechanical damages of the specimen, unfortunately, the geometry of the hypothetical exit points B and C cannot be reconstructed.

4. Possibility of impregnation of the leakage path

Based on the previous chapter, the realistic geometry of the leaky cavity system is as follows: the volume of the cavity system is 12 mm^3, the minimum internal radius of the cavity system is 10 μm, the length of the capillary-like section running out to the surface is 50-150 μm, with a radius of 1-4 μm. Based on preliminary calculations, the vacuuming and saturation of the leakage cavity system is fully achieved when the internal radius of the cavity system, assumed to be cylindrical, exceeds 7 μm. The principle and details of the relevant calculation are described in the previous study [5]. The developed model and computational procedure are suitable for monitoring the pressure and saturation conditions while taking into account the realistic dimensions of the cavity system. In the model calculation, it is assumed that the capillary geometry at both ends of the leaky cavity is the same, so that it is sufficient to consider the half of the geometry of the cavity system. The results of the calculation are shown in the diagrams in Figure 6 for a cavity system with an overall volume of 6 mm^3 (half of the total volume), connected to the surface by a capillary-like section of length 100 μm and radius less than 4 μm.

a) SEM image of the thread forming profile

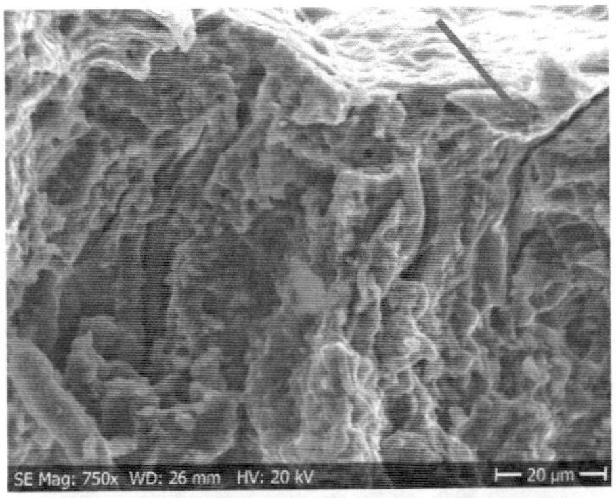

b) exit point of the leak channel

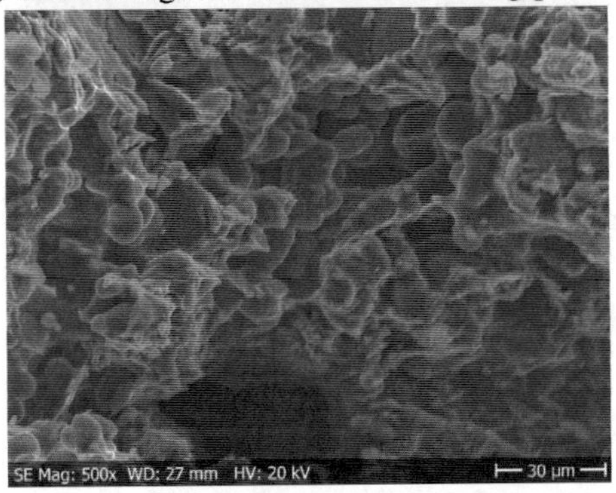

c)

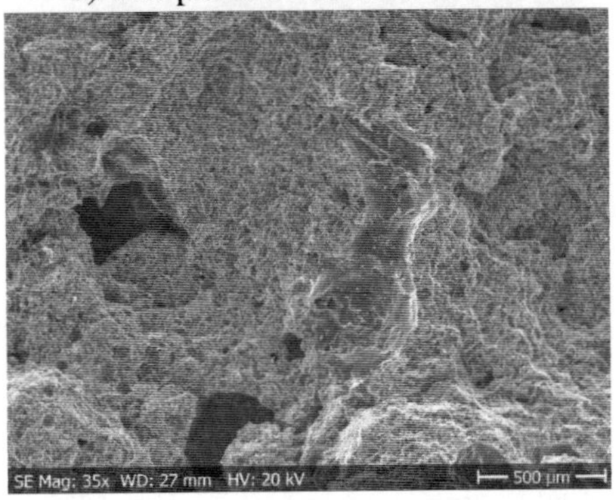

d)

Figure 5 Scanning electron micrographs of the characteristic parts of the surface of the array

The figures show the *p0* pressure of the atmosphere surrounding the part (blue line), the *p1* pressure in the cavity (red line), the *delta p* pressure difference between them (black line) and the degree of saturation of the cavity with resin (green line, in percent) during the vacuum (0-10 min) and saturation (10-16 min) stages of the impregnation. Each figure also shows the capillary pressure value for the given capillary radius, indicated by a dashed line. Figure 6.a shows the calculation results for the case of the 4 µm capillary radius. It can be seen that in this case both vacuuming and saturation are successful, so by the end of the impregnation process the cavity has reached 99% saturation with resin. If the capillary radius is 3 (Figure 6.b) or 2 µm (Figure 6.c), the saturation process cannot be fully achieved, so the cavity saturation is only 40 and 5 %, respectively. If the capillary radius is only 1 µm, even the vacuuming process cannot be considered to be successful, as only 32 kPa pressure is achieved in the available 10 minutes. Because of this, the saturation of the cavity is minimal (less than 1 %), i.e. the cavity remains permeable to gases at the end of the impregnation. Also, the capillary section itself is not saturated with resin, as the pressure difference between the outer and inner spaces (delta p) is greater than the capillary pressure.

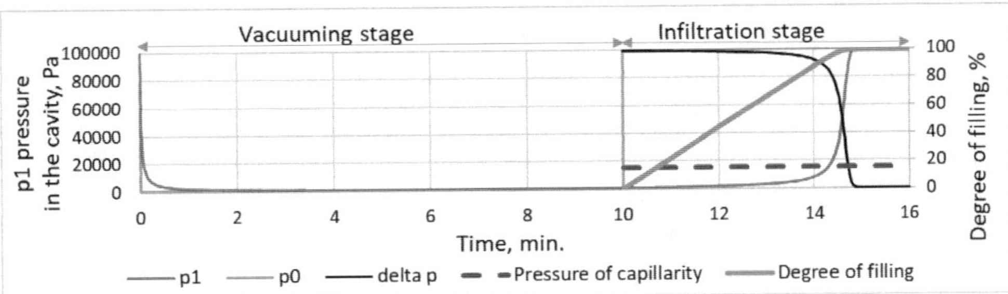

a) Capillary radius 4 µm

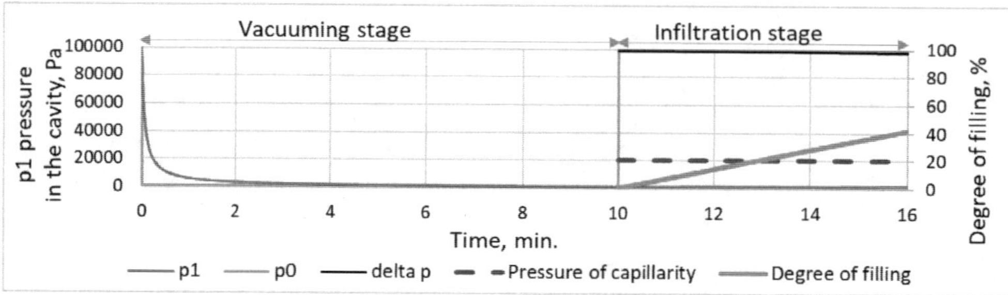

b) Capillary radius 3 µm

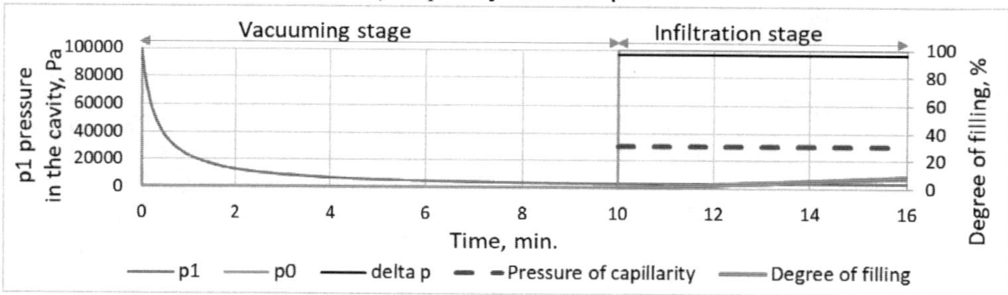

c) Capillary radius 2 µm

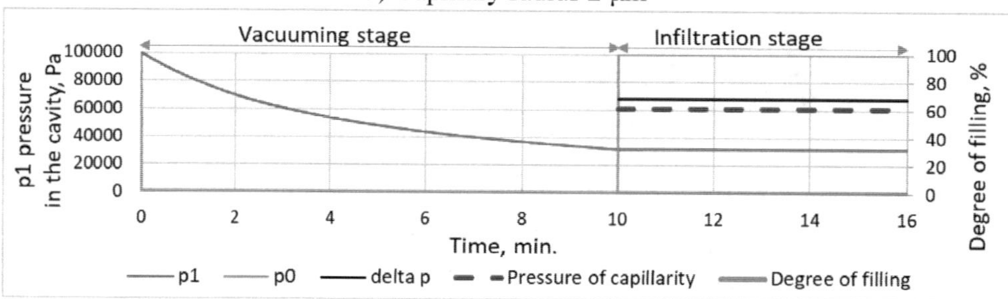

d) Capillary radius 1 µm

Figure 6 Vacuuming and filling with resin of a cavity with a volume of 6 mm^3 connected to the surface by a capillary section of 100 µm length for different capillary radius sizes

5. Summary

Experience has shown that high quality aluminum alloy die cast products manufactured with a gas tightness requirement can show leakage even after the impregnation operation. The machining operations can potentially expose capillaries, while the thread forming technology may be able to reduce or eliminate the size of capillaries running out onto the surface. Based on a non-destructive and destructive examination of a proven leaking casting, the suspected leakage pathway was identified and its geometric characteristics were determined. The total volume of the cavity system of the leakage pathway is between 10-20 mm^3, with an average internal diameter of the cavity system of about 180 µm. It is assumed that a capillary-like section of 50-150 µm length is formed in the near-surface sections of the cavity system due to the deformation process caused by machining. If the inner diameter of the capillary section is less than 6-7 µm, it is calculated that the cavity system is not or only partially impregnated, i.e. the cavity system can remain permeable to gases even after impregnation.

Acknowledgments

This research was funded by the 2020-1.1.2-PIACI-KFI-2020-00129 project. The authors acknowledge the financial support of this work by Hungarian-Japanese bilateral project (2019-2.1.11-TÉT-2020-00204.)

References

[1] Young, P., *Impregnation of Porous Castings.* Foundry Trade Journal, 1974. **136**(2993): p. 439-431.

[2] Holm, V.C.F., *A Study of Resinous Sealants for Porous Metal Casting,* Research Paper, U. S. Department of Commerce National Bureau of Standards, 1946, **Volume 37**.

[3] Lloyd, G., *Finding a Solution to the Eternal Problem of Porosity in Casting.* Die Casting Engineer, 2012. **40**.

[4] https://www.ultraseal-impregnation.com/wp-content/uploads/2017/08/Ultraseal-Data-Sheet-v1-Rexeal-100-Sealant-v3.pdf

[5] *http://everyspec.com/MIL-SPECS/MIL-SPECS-MIL-I/MIL-I-17563C_28872/.* Military specification impregnants for cast or powder metal components.

[6] Reger M , H.R., Fabian ER, Reti T, *Modelling the impregnation of a gastight casting.* International Journal of Metalcasting, 2024. https://doi.org/10.1007/s40962-024-01272-1

[7] Campanelli, S., et al, *Study of the aging treatment on selective laser melted maraging 300 steel.* Materials Research Express, 2019. **6**(6): p. 066580. https://doi.org/10.1088/2053-1591/ab0c6e

[8] Akarca, S.S., *Modeling of subsurface deformation and damage in an aluminum-silicon alloy subjected to sliding contact.* 2008.

[9] Liang, X., et al., *Tool wear-induced microstructure evolution in localized deformation layer of machined Ti-6Al-4V.* Journal of Materials Science, 2020. **55**(8): p. 3636-3651. https://doi.org/10.1007/s10853-019-04214-z

[10] Simon, M. and C. Sauerwein. *Quality control of light metal castings by 3D computed tomography.* in *15th world conference on non-destructive testing.* 2000.

[11] Jaques, V.A., et al., *Review of porosity uncertainty estimation methods in computed tomography dataset.* Measurement Science and Technology, 2021. **32**(12): p. 122001. https://doi.org/10.1088/1361-6501/ac1b40

[12] Carmignato, S., *Computed tomography as a promising solution for industrial quality control and inspection of castings.* Metallurgical Science and Tecnology, 2012. **30**(1).

[13] Garb, C., et al., *Statistical analysis of micropore size distributions in Al–Si castings evaluated by X-ray computed tomography.* International Journal of Materials Research, 2018. **109**(10): p. 889-899. https://doi.org/10.3139/146.111685

[14] Weidt, M., et al., *Prediction of porosity characteristics of aluminium castings based on X-ray CT measurements.* International Journal of Cast Metals Research, 2018. **31**(5): p. 289-307. https://doi.org/10.1080/13640461.2018.1467105

[15] Nicoletto, G., R. Konečná, and S. Fintova, *Characterization of microshrinkage casting defects of Al–Si alloys by X-ray computed tomography and metallography.* International journal of fatigue, 2012. **41**: p. 39-46. https://doi.org/10.1016/j.ijfatigue.2012.01.006

[16] Nourian-Avval, A. and A. Fatemi, *Characterization and analysis of porosities in high pressure die cast aluminum by using metallography, x-ray radiography, and micro-computed tomography.* Materials, 2020. **13**(14): p. 3068. https://doi.org/10.3390/ma13143068

[17] Roy, N., A. Samuel, and F. Samuel, *Porosity formation in Al-9 Wt pct Si-3 Wt pct Cu alloy systems: Metallographic observations.* Metallurgical and Materials transactions A, 1996. **27**(2): p. 415-429. https://doi.org/10.1007/BF02648419

[18] Ran, G. and J.E. Zhou. *Metallographic characterization of porosity in a cast aluminum alloy A356-T6.* in *Materials science forum.* 2007. Trans Tech Publ. https://doi.org/10.4028/www.scientific.net/MSF.546-549.989

[19] Felberbaum, M., *Porosity in Aluminum Alloys.* 2010, EPFL. https://doi.org/10.5075/epfl-thesis-4639

IOP Conf. Series: Materials Science and Engineering **1313** (2024) 012008 doi:10.1088/1757-899X/1313/1/012008

Thermomechanical-chemical devulcanization of ground tire rubber

Á Görbe, A Kohári, I Z Halász-Kutasi, T Bárány
Budapest University of Technology and Economics, Faculty of Mechanical Engineering, Department of Polymer Engineering, H-1111 Budapest, Műegyetem rkp. 3., Hungary
E-mail: barany@pt.bme.hu

Abstract. We developed a thermomechanical devulcanization process of ground tire rubber (GTR) by incorporating chemicals to achieve better devulcanization. The samples were devulcanized in a twin-screw extruder with the help of three types of chemicals that aid devulcanization: N-(cyclohexylthio) phthalimide (PVI), dibenzamid diphenyl disulfide (DBD) and dialkyl-pentasulfide (DPS, commercialized as Aktiplast 79). We characterized the resulting devulcanizates by Soxhlet-extraction, and produced vulcanized sheets, which we characterized mechanically and examined the vulcanization process. We found that all chemicals facilitated better devulcanization and the devulcanizate had a more rubber-like behavior during mixing. Vulcanization was also slightly improved and reversion was reduced. When revulcanized, the rubber had improved strain at break and tear strength.

Introduction

The upcycling of elastomers is one of the most critical problems in the recycling of polymers [1]. This is due to the structure of elastomers: their high elasticity is provided by crosslinks between their molecule chains. These crosslinks, however, make them hard to recycle as they hinder reversible meltability. The most pressing issue of elastomer recycling is the recycling of the tire rubbers. Tire rubbers are composite materials consisting of metallic and textile reinforcements and different types of rubbers. First, the elastomeric part must be ground, producing ground tire rubber (GTR). This GTR is easy to handle, and it can be used to toughen different kinds of polymers, for example, thermoplastics [2-5], thermosets [6, 7], and rubbers as well [8-11]. The problem with this kind of use is that the GTR is not compatible with the matrix, so these materials cannot be high-performance materials. While this use of GTR is viable, it cannot be considered optimal as it cannot handle the increasing mass of rubber waste. A better way of recycling is devulcanization and reclamation, both of which aim to break up the crosslinked structure of rubbers, thus enabling primary recycling [12]. Devulcanization achieves this goal by selective crosslink scission; reclamation is also accompanied by significant chain scission [13]. There are several ways of devulcanization, the most widespread of which are the thermomechanical [14-16], thermochemical [17, 18] and microwave [19, 20] methods. The most straightforward of these is thermomechanical devulcanization because it is continuous, highly productive, and does not necessarily involve chemical agents [15]. Thermomechanical devulcanization is carried out mainly in a twin-screw extruder, where combining heat and shear produces enough energy to break crosslinks. [21]. Heat and shear can be altered easily in the extruder as extruder screws are modular and thus easily customizable. Thermomechanical devulcanization can be aided with chemical additives that help break crosslinks and they also influence recombination [13]. The degree of devulcanization can be rated by soluble content and cross-link density. Soluble fraction can be used to quantify the degree of devulcanization since devulcanizates are more soluble than crosslinked elastomers [14, 22, 23].

The main disadvantage of devulcanizates is that if they are revulcanized, their properties are inferior to those of primary rubbers. Researchers also found that revulcanizing is significantly faster than in the

case of conventional rubbers due to the accelerator residue remaining after devulcanization [24]. This fast revulcanization is also facilitated by residual carbon black as it has oxygen, nitrogen, and sulphur on its surface [25]. Fast revulcanization might also result from devulcanization itself, as reduced cross-linking density increases polymeric chain mobility, leading to more effective collisions among molecules and a more rapid revulcanization reaction [26].
In this study, our aim was to improve thermomechanical devulcanization by incorporating different types of devulcanization aids. Additionally, we aimed to improve the consistency of the devulcanizate, to create a more rubber-like material that can be processed more easily by the rubber industry.

Materials and methods

Materials

We used ground tire rubber (GTR) from AquaJet Ltd. (Budapest, Hungary), produced by water jet milling with a particle size under 0.4 mm. The pellets were made from truck tire treads, they were clean and contained only a small amount of contaminants.

We used three types of devulcanization agents:
- PVI: N-(cyclohexylthio)phthalimide
- DBD: Dibenzamid diphenyl disulfide
- Aktiplast 79: Dialkyl-pentasulfide

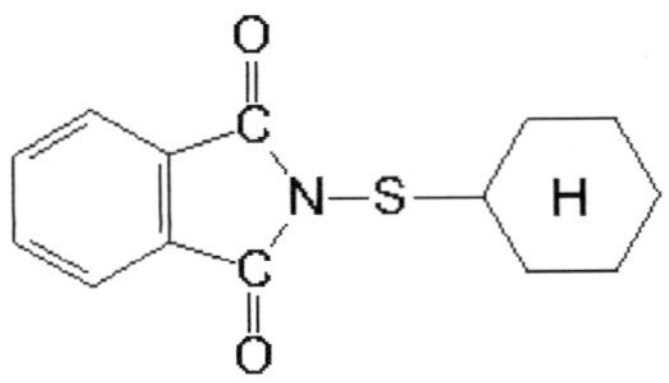

a)

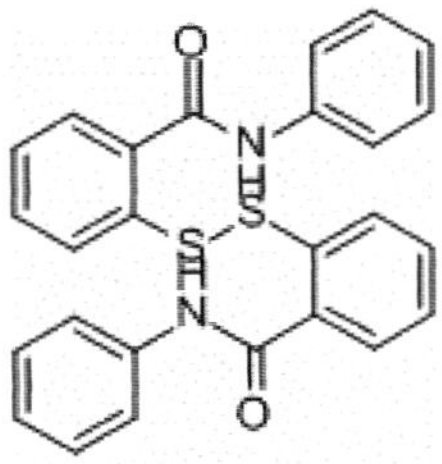

b)

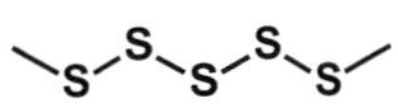

c)

Figure 1.: The chemical composition of the devulcanization aids (a): PVI, b): DBD, c): Aktiplast 79).

PVI is used in the rubber industry as a retarder, DBD is a known peptizer used in the treatment of natural rubber, and Aktiplast 79 is a reclamation agent for natural and synthetic rubbers. We formulated six recipes using these materials (Table 1).

Table 1.: The compounds formulated for devulcanization.

Content	Amount (phr)					
	1	2	3	4	5	6
GTR	100	100	100	100	100	100
PVI	0	2	0	0	1	1
DBD	0	0	2	0	1	0
Aktiplast 79	0	0	0	2	0	1

We constructed a simple recipe for revulcanizing the devulcanized GTRs (dGTRs) (Table 2). Zinc oxide (Werco Metal (Zlatna, Romania)) and stearic acid (Oleon (Ertvelde, Belgium)) were used as activators,

CBS (*N-Cyclohexyl-2-benzothiazolesulfenamide)* (Rhein Chemie (Mannheim, Germany)) was used as accelerator, and sulphur (Ningbo Actmix Polymer (Zhejiang, China)) was used for cross-linking.

Table 2.: Recipes for revulcanization.

Content	Amount (phr)
dGTR (with devulcanization aids)	100
Zinc oxide	5
Stearic acid	2
CBS	1.5
Sulphur	1.5

Methods

The devulcanizing agents were mixed with GTR in a Henschel FM/A10 Fluid Mixer (Henschel (Kassen, Germany) at 1000 rpm for 5 min.

Devulcanization was carried out with a twin-screw extruder LTE 26-44 (Labtech Engineering Co., Ltd., Samutprakarn, Thailand) with unidirectionally rotating screws at a barrel temperature of 190 °C and a speed of 60 rpm.

The vulcanizing agents were mixed to dGTR with a Brabender Lab-Station internal mixer (Brabender GmbH & Co. KG, Duisburg, Germany). We used the Intermix (Mixer type 350 SX) mixing module. Mixing was performed at a speed of 40 rpm for 6 min in a chamber at 50 °C with the chamber filled approximately 60%.

The dGTR samples were characterized by Soxhlet extraction and compared based on soluble content, which allowed a comparison of their degree of devulcanization. Measurements were performed in a Soxhlet extractor with a hot toluene solvent for 18 h and then dried for 12 h. The soluble solid content is given by the formula below:

$$Sol\ content\ (\%) = \left(1 - \frac{M_f}{M_i}\right) \cdot 100$$

where M_i and M_f stand for sample mass before and after extraction.

We used a D-RPA 3000 moving die rheometer (MonTech, Buchen, Germany) to record the vulcanization curves of dGTR samples and determine the vulcanization times. Measurements were performed at 160 and 180 °C.

Then the revulcanized rubbers were tested. Square sheets of 160x160 mm and 2 mm thickness were pressed with a Teach-Line Platen Press 200E (Dr. Collin GmbH, Munich, Germany) hydraulic press at 180 °C for the typical vulcanization time t_{90} + 30 seconds. This ensured that the whole volume of the sample was vulcanized. The pressure was 16 bar.

We performed tensile and tear tests to characterize the revulcanized rubbers using a Zwick Z005 (Zwick, Ulm, Germany) universal tensile tester with a speed of 500 mm/min. Tensile strength and elongation at break were calculated with the following equations:

$$\sigma_B = \frac{F_B}{A_0}$$

$$\varepsilon_B = \frac{L_B - L_0}{L_0} \cdot 100$$

where σ_B is tensile strength (MPa), F_B is the force measured at break (N), A_0 is the initial cross-section of the specimens (mm^2), ε_B is elongation at break (%), L_B is the length measured at break (mm), and L_0 is the initial gauge length of the specimen (mm). Tear tests were performed on standardized specimens with a 1 mm nick. We characterized the samples by tear strength, which can be calculated in the following way:

$$T_s = \frac{F_m}{a_0}$$

where T_s is tear force (N/mm), F_m is the maximal force (N), and a_0 is the initial thickness of the samples (mm).

Results and discussion

Soluble content

The soluble content of the samples (Table 3) showed that the additives incorporated into the GTR were almost completely dissolved without devulcanization, as the soluble content before devulcanization typically increased by 2%, which was equal to the amount of additives. This suggests that elevated temperatures are indeed required for the mechanism of action of the devulcanization agents. After devulcanization, the soluble content of all devulcanizates increased significantly compared to the reference dGTR, so the devulcanization agents facilitated devulcanization. This effect was most pronounced with DBD, which proved to be the most effective additive, while the least increase was produced by PVI. This can be explained with the fact that DBD decreases the molecular weight of crosslinked elastomers, and this facilitates a better degree of devulcanization.

Table 3.: Soluble content.

Sample (before devulcanization)	Sol content (%)	Sample (after devulcanization)	Sol content (%)
GTR_0	10.8 ± 1.0	**$dGTR_0$**	13.3 ± 0.7
GTR_{PVI}	12.3 ± 1.0	**$dGTR_{PVI}$**	17.0 ± 1.0
GTR_{DBD}	12.7 ± 0.3	**$dGTR_{DBD}$**	20.5 ± 0.9
$GTR_{Aktiplast\ 79}$	12.2 ± 0.6	**$dGTR_{Aktiplast\ 79}$**	19.0 ± 0.6
$GTR_{PVI+DBD}$	12.5 ± 0.9	**$dGTR_{PVI+DBD}$**	18.6 ± 0.1
$GTR_{PVI+Aktiplast\ 79}$	11.9 ± 0.8	**$dGTR_{PVI+Aktiplast\ 79}$**	17.0 ± 0.7

The increase in devulcanization was also apparent from the consistency of the dGTRs: using devulcanization aids produced a rubber-like devulcanizate instead of rubber granules (Figure 2).

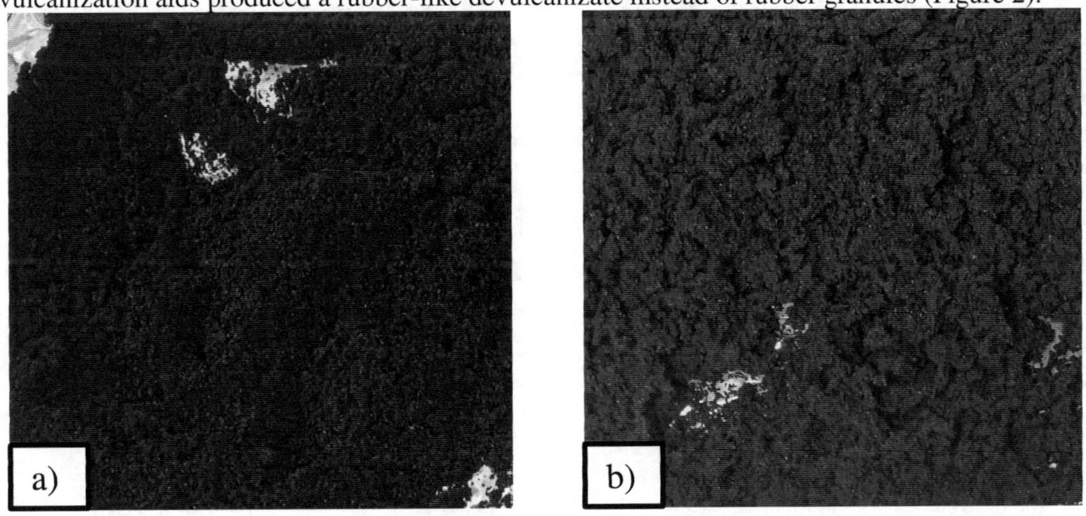

Figure 2.: Comparison of the consistency of devulcanizates (a): without devulcanization aids, b): with devulcanization aids).

IOP Conf. Series: Materials Science and Engineering **1313** (2024) 012008 doi:10.1088/1757-899X/1313/1/012008

Mixing

During the preparation of the rubber mixtures, we registered the torque and temperature of the mixture, and drew conclusions from them. The mixing curves (Figure 3) indicate that all devulcanization agents increased the torque at the end of mixing. This can be attributed to a more rubber-like behavior, since rubbers have higher viscosity than other plastics, and this can increase mixing torque. Increasing torque also results in growing temperatures at the end of mixing. This effect is most pronounced in the case of Aktiplast 79, which seems to increase viscosity of the devulcanizates, and this also showed in the presence of dGTRs: while all others came out of the extruder in a powdery form, the Aktiplast 79 helped to create a more consistent material.

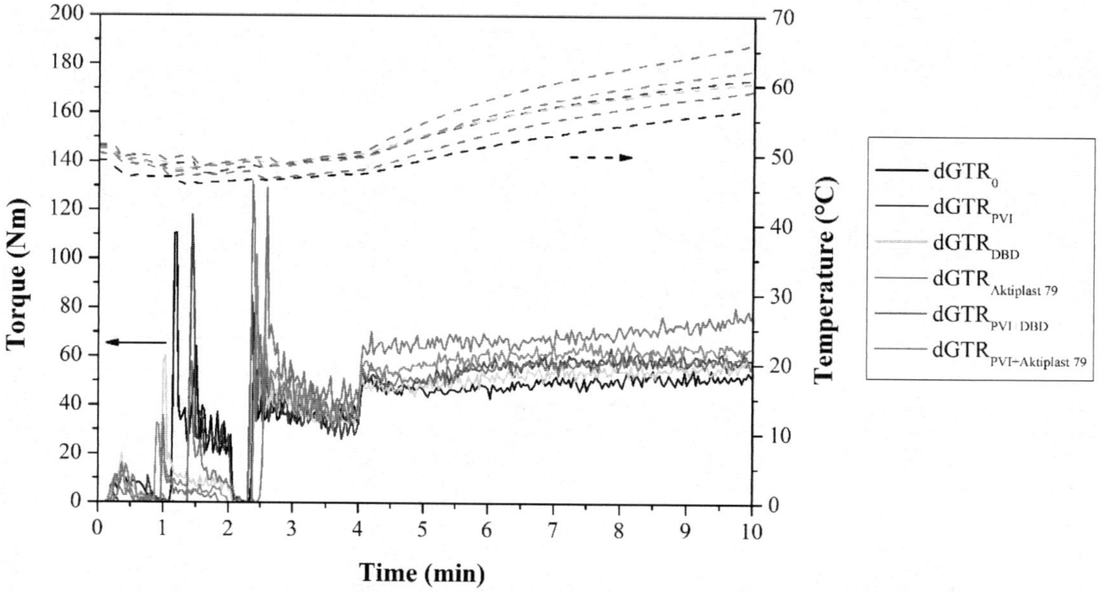

Figure 3.: The recorded plastograms.

Vulcanization

In our analysis of the vulcanization process (Figure 4, Table 4), we observed a consistent decrease in torque with all additives. This decrease suggests that these additives produce a softer behavior in the material, which indicates shorter molecules. Notably, Aktiplast 79 produced the most substantial reduction in torque, showing its significant impact.

However, despite the use of these additives, there was no significant improvement in the vulcanization times of dGTR blends. Only a marginal deceleration was noticed with the application of PVI, failing to alter the inherently rapid vulcanization process notably.

Furthermore, the combined use of PVI with other additives like DBD or Aktiplast 79 did not yield any synergistic effects on the vulcanization process.

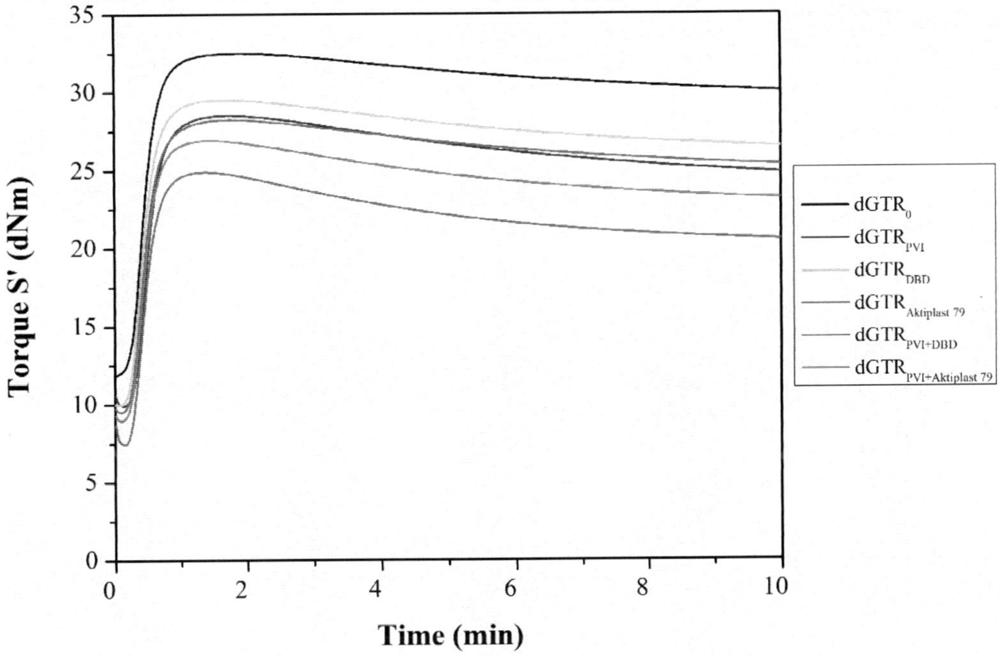

Figure 4.: Vulcanization curves recorded at 180 °C.

Table 4.: Characteristics of the vulcanization process.

Samples	S'_{min} (dNm)	S'_{max} (dNm)	t_{10} (min)	t_{30} (min)	t_{50} (min)	t_{70} (min)	t_{90} (min)
$dGTR_0$	12.07	32.52	0.26	0.36	0.44	0.54	0.75
$dGTR_{PVI}$	9.92	28.52	0.31	0.41	0.49	0.59	0.81
$dGTR_{DBD}$	10.03	29.52	0.25	0.35	0.43	0.53	0.73
$dGTR_{Aktiplast\ 79}$	7.46	24.91	0.28	0.37	0.45	0.54	0.73
$dGTR_{PVI+DBD}$	9.53	28.23	0.27	0.37	0.44	0.55	0.76
$dGTR_{PVI+Aktiplast\ 79}$	8.98	26.92	0.27	0.36	0.44	0.54	0.72

Tensile tests

The tensile test results (Figure 5, Table 5) showed that the devulcanizing aids led to a noteworthy increase in elongation at break for the revulcanized rubbers. This enhancement, however, came at the cost of reduced tensile strength. Among the tested chemicals, Aktiplast 79 improved elongation at break most, which can be attributed to the enhanced devulcanization produced by Aktiplast 79. Interestingly, there was no synergistic effect observed with the simultaneous use of PVI and Aktiplast 79.

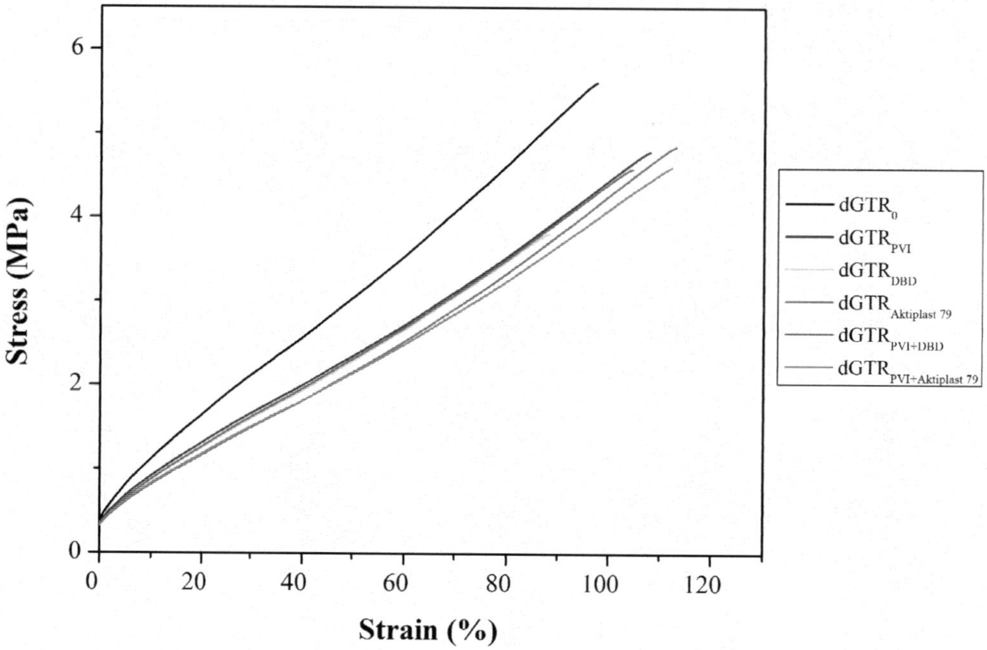

Figure 5.: Typical tensile curves.

Table 5.: Mechanical properties of the revulcanized rubbers.

Samples	Maximal stress (MPa)	Strain at break (%)	Tear strength (N/mm)
dGTR$_0$	5.4 ± 0.4	98.4 ± 1.5	4.4 ± 0.2
dGTR$_{PVI}$	4.7 ± 0.3	109.6 ± 6.6	5.2 ± 0.2
dGTR$_{DBD}$	3.9 ± 0.4	89.7 ± 5.0	5.4 ± 0.6
dGTR$_{Aktiplast\ 79}$	4.8 ± 0.1	114.2 ± 8.2	6.1 ± 0.8
dGTR$_{PVI+DBD}$	4.6 ± 0.4	108.1 ± 7.5	4.9 ± 0.2
dGTR$_{PVI+Aktiplast\ 79}$	4.4 ± 0.9	109.1 ± 19.9	5.0 ± 0.1

Tear tests

The results of the tear tests (Figure 6, Table 5) clearly show that the devulcanization aids increased tear propagation strength, and improved resistance against crack propagation in all cases. This improvement was most striking with Aktiplast 79, there was no synergistic effect when it was used with PVI. This behavior was most likely in connection with the higher rate of devulcanization, which yielded better revulcanizates as well.

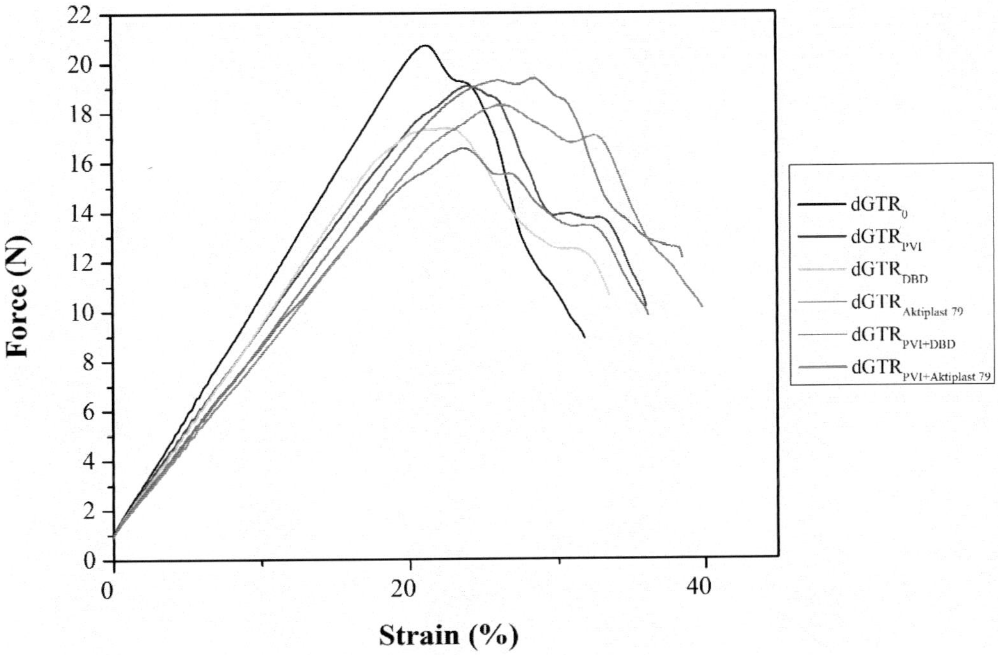

Figure 6.: Typical tear curves.

Conclusions

We presented a way of enhancing thermomechanical devulcanization by using devulcanization chemicals, PVI, DBD, and Aktiplast 79, and examined devulcanization, mixing, and revulcanization. We also characterized the revulcanized rubbers with mechanical tests and concluded that all of the chemicals increased the degree of devulcanization, and the revulcanized rubbers were also improved when devulcanized in the presence of these chemicals. The greatest improvement was observed with Aktiplast 79, which also produced a rubber-like devulcanizate instead of rubber granules.

Acknowledgments

The project supported by the Doctoral Excellence Fellowship Programme (DCEP) is funded by the National Research Development and Innovation Fund of the Ministry of Culture and Innovation and the Budapest University of Technology and Economics under a grant agreement with the National Research, Development and Innovation Office. This work was supported by the National Research, Development and Innovation Office, Hungary (K128268).
The authors would like to express their gratitude towards Dr. Tibor Nagy for his advice and valuable help.

References

[1] Valentini F., Pegoretti A.: End-of-life options of tyres. A review. Advanced Industrial and Engineering Polymer Research, **5**, 203-213 (2022).
[2] Karger-Kocsis J., Mészáros L., Bárány T.: Ground tyre rubber (gtr) in thermoplastics, thermosets, and rubbers. Journal of Materials Science, **48**, 1-38 (2012).
[3] Kiss L., Simon D. Á., Petrény R., Kocsis D., Bárány T., Mészáros L.: Ground tire rubber filled low-density polyethylene: The effect of particle size. Advanced Industrial and Engineering Polymer Research, **5**, 12-17 (2022).

[4] Kumar C. R., Fuhrmann I., Karger-Kocsis J.: Ldpe-based thermoplastic elastomers containing ground tire rubber with and without dynamic curing. Polymer Degradation and Stability, **76**, 137-144 (2002).
[5] Ramarad S., Khalid M., Ratnam C., Chuah A. L., Rashmi W.: Waste tire rubber in polymer blends: A review on the evolution, properties and future. Progress in Materials Science, **72**, 100-140 (2015).
[6] Hejna A., Korol J., Przybysz-Romatowska M., Zedler L., Chmielnicki B., Formela K.: Waste tire rubber as low-cost and environmentally-friendly modifier in thermoset polymers - a review. Waste Management, **108**, 106-118 (2020).
[7] Taurino R., Bondioli F., Messori M.: Use of different kinds of waste in the construction of new polymer composites: Review. Materials Today Sustainability, **21**, (2023).
[8] Candau N., Leblanc R., Maspoch M. L.: A comparison of the mechanical behaviour of natural rubber-based blends using waste rubber particles obtained by cryogrinding and high-shear mixing. Express Polymer Letters, **17**, 1135-1153 (2023).
[9] Pastor L. E. A., Carrero K. C. N., González M., Araujo-Morera J., Peters G., Pastor J. M., Santana M. H.: Life cycle assessment applied to a self-healing elastomer filled with ground tire rubber. Journal of Cleaner Production, **419**, (2023).
[10] Fazli A., Rodrigue D.: Recycling waste tires into ground tire rubber (gtr)/rubber compounds: A review. Journal of Composites Science, **4**, 103 (2020).
[11] Karabork: Investigation of the mechanical, tribological and corrosive properties of epoxy composite coatings reinforced with recycled waste tire products. Express Polymer Letters, **16**, 1114-1127 (2022).
[12] Saputra R., Walvekar R., Khalid M., Mubarak N. M., Sillanpää M.: Current progress in waste tire rubber devulcanization. Chemosphere, **265**, 129033 (2021).
[13] Bockstal L., Berchem T., Schmetz Q., Richel A.: Devulcanisation and reclaiming of tires and rubber by physical and chemical processes: A review. Journal of Cleaner Production, **236**, 117574 (2019).
[14] Simon D. Á., Barany T.: Effective thermomechanical devulcanization of ground tire rubber with a co-rotating twin-screw extruder. Polymer Degradation and Stability, **190**, 109626 (2021).
[15] Seghar S., Asaro L., Rolland-Monnet M., Hocine N. A.: Thermo-mechanical devulcanization and recycling of rubber industry waste. Resources, Conservation and Recycling, **144**, 180-186 (2019).
[16] Asaro L., Gratton M., Seghar S., Hocine N. A.: Recycling of rubber wastes by devulcanization. Resources, Conservation and Recycling, **133**, 250-262 (2018).
[17] Ghorai S., Bhunia S., Roy M., De D.: Mechanochemical devulcanization of natural rubber vulcanizate by dual function disulfide chemicals. Polymer Degradation and Stability, **129**, 34-46 (2016).
[18] Gumede J. I., Hlangothi B. G., Woolard C. D., Hlangothi S. P.: Organic chemical devulcanization of rubber vulcanizates in supercritical carbon dioxide and associated less eco-unfriendly approaches: A review. Waste Management & Research, **40**, 490-503 (2022).
[19] Simon D. Á., Bárány T.: Microwave devulcanization of ground tire rubber and its improved utilization in natural rubber compounds. ACS Sustainable Chemistry & Engineering, **11**, 1797-1808 (2023).
[20] Garcia P., De Sousa F., De Lima J., Cruz S., Scuracchio C.: Devulcanization of ground tire rubber: Physical and chemical changes after different microwave exposure times. Express Polymer Letters, **9**, (2015).
[21] Formela K., Cysewska M.: Efficiency of thermomechanical reclaiming of ground tire rubber conducted in counter-rotating and co-rotating twin screw extruder. Polimery, **59**, 231-238 (2014).
[22] Horikx M.: Chain scissions in a polymer network. Journal of Polymer Science, **19**, 445-454 (1956).
[23] Formela K., Cysewska M., Haponiuk J. T.: Thermomechanical reclaiming of ground tire rubber via extrusion at low temperature: Efficiency and limits. Journal of Vinyl and Additive Technology, **22**, 213-221 (2016).
[24] Ismail H., Ishak S., Hamid Z.: Effect of blend ratio on cure characteristics, tensile properties, thermal and swelling properties of mica-filled (ethylene-propylene-diene monomer)/(recycled ethylene-propylene-diene monomer)(epdm/r-epdm) blends. Journal of Vinyl and Additive Technology, **21**, 1-6 (2015).
[25] Roychoudhury A., De P.: Reinforcement of epoxidized natural rubber by carbon black: Effect of surface oxidation of carbon black particles. Journal of applied polymer science, **50**, 181-186 (1993).

[26] de Sousa F. D. B., Zanchet A., Ornaghi Júnior H. L., Ornaghi F. G.: Revulcanization kinetics of waste tire rubber devulcanized by microwaves: Challenges in getting recycled tire rubber for technical application. ACS Sustainable Chemistry & Engineering, **7**, 15413-15426 (2019).

IOP Conf. Series: Materials Science and Engineering **1313** (2024) 012009 doi:10.1088/1757-899X/1313/1/012009

Some considerations on the contradictions of the theory of Frank–Read sources

János Dobránszky[1] **and Balázs Verő**[3]

[1] HUN-REN–BME Research Group for Composite Science and Technology, 1111 Budapest, Műegyetem rakpart 3, Hungary
[2] University of Dunaújváros, 2800 Dunaújváros, Táncsics Mihály utca 1/a, Hungary

E.mail: Dobranszky.Janos@eik.bme.hu

Abstract. The basis of the theoretical explanation of the plastic deformation of metals is the sliding of dislocations and the dislocation multiplication that takes place during the deformation. The best-known interpretation of this process is based on the operation of Frank-Read sources. The article tries to shed light on aspects and theorems that have been known for decades, which draw attention to quite a few contradictions in the theory of Frank-Read sources.

1. Introduction

The theoretical explanation of the plastic deformation of solid bodies, especially metals and alloys, has been one of the most intensively discussed topics in material science in today's sense since the very beginning of 1900. From the beginning, the development of the theories of plasticity proceeded along several paths, according to which the analysis is carried out with the partially overlapping but still independent tools of metallurgical technology, continuum mechanics, mechanical metallurgy, physical metallurgy or metal physics.

Hungarian scientists achieved significant results in each field of science. In Hungary, *Sándor Rejtő* (1907, 1914, 1925) [1, 2, 3], *Vilmos Misángyi* (1908) [4], *István Gállik* (1909) [5], *József Verő* (1931) [6] and *Frigyes Geleji* (1942) [7] were the pioneers. Among Hungarians working abroad, the research results of *Tódor Kármán*'s Habilitation thesis in Göttingen (1910) [8], and then *Árpád Lajos Nádai* (1927) [9] *Egon Orován* (1934) [10] and *Mihály Polányi* (1934) [11] became of decisive importance. *Orowan*, *Polanyi* – and *G.I. Taylor*, who also published his insights in 1934 [12] – are today considered by the world as the founders of the dislocation theory of the plastic deformation of crystalline materials: the idea of the propagation of dislocations as the mechanism of slip in crystals was first put forward independently in 1934 by *Orowan*, *Polanyi* and *Taylor* [13].

After the publication of the Orowan–Polányi–Taylor theorems for the interpretation of the plastic deformation of metals by dislocation slip, explanations were sought for many years as to how the number of dislocations can multiply during the plastic deformation process. There have been quite a few theories. Among these, the theory of *Frank* and *Read* (1950) was born [14], which at the beginning is often noted as a particularly rare case, but when ten years later (without scale lines) photographs began to be published [15], they have already been called "dramatic proof" of the existence of the Frank–Read sources. Nevertheless, some specialist books still only mentioned it as a rare phenomenon.

To date, the explanation of dislocation multiplication by the operation of Frank–Read sources has had an exceptional career. At the turn of the millennium, so approx. after fifty years, a considerable part of the published technical books and articles talk almost exclusively about this mechanism. There

IOP Conf. Series: Materials Science and Engineering **1313** (2024) 012009 doi:10.1088/1757-899X/1313/1/012009

are also those who believe that the discovery of the Frank–Read sources was a decisive step in the application of the theory of dislocations to plasticity [16].

It is only possible to find other theories of dislocation multiplication by digging deep into the past, namely, e.g. Schmid–Boas (1935), Peierls–Mott–Nabarro (1947), Cottrell–Bilby (1949), Bardeen–Herring (1950), Peach–Kohler (1950), Weertman–Weertman (1964), to the Holt (1970), Hirth–Lothe (1992) or de Koning–Wei Caj–Bulatov (2003) theories.

We believe that the secret of the successful spread of the Frank–Read theory lies mainly in the fact that a very expressive planar representation could be associated with it; Figure 1 shows one of these.

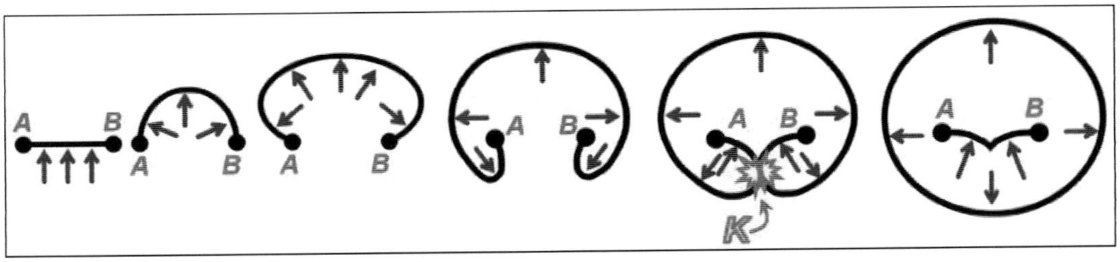

Figure 1. Planar schema of the operation of the Frank–Read source; points A and B are the sections of a spatial dislocation that fall into the slip plane, and point K is where the two points of the buckling dislocation line meet. The blue arrows indicate the displacement direction of the given point of the dislocation line and the direction of the shear stress acting at the given point.

The explanations given about the origin of the screw dislocation are logically correct. In contrast, the usual discussion of the formation of edge dislocations lacks any reality. In general, we can read that take a block of crystalline material, cut it in half along one of its atomic planes, and insert a half-plane of extra atoms here, then glue it together and let the atoms relax elastically; this is illustrated in Figure 2. The last atomic row of the half-plane formed by the excess atoms forms the edge dislocation line, which can move perpendicular to the half-plane during plastic deformation. Although the described imaginary formation of the edge dislocation characterizes the geometrical situation, it does not say where the atoms of the extra half-plane come from, nor how these half-planes multiply.

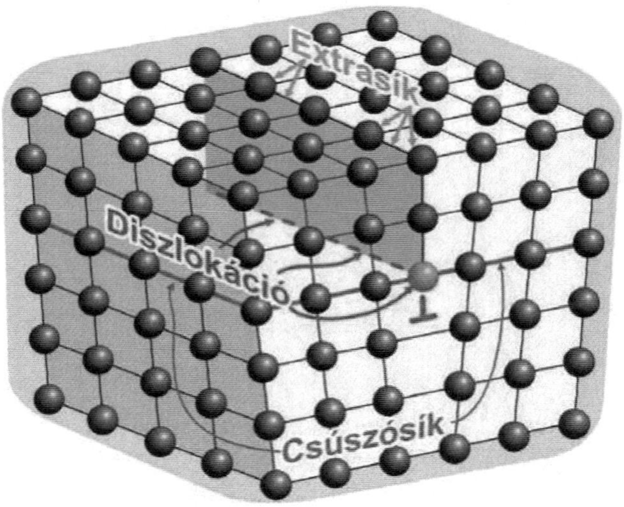

Figure 2. The usual representation of the edge dislocation (red) with the inserted extra half-plane (orange) and the slip plane (blue)

However, an edge dislocation is required according to the Frank–Read mechanism, which, as mentioned, has become the most well-known theory of dislocation multiplication, even though a few decades ago the relevant literature classified it as extremely rare. This is also a sufficient reason to state: the theory of Frank–Read sources is perhaps the most controversial in the given topic. In the following, these contradictions are the focus of our analysis.

2. The inconsistency of the theory of Frank–Read sources

Frank–Read means the source, itself, the dislocation section in the slip plane, fixed at its two endpoints. This dislocation section bends under the influence of the external sliding stress, until the stress resulting from the external load balances with the component in the direction corresponding to the line stress of the dislocation. In equilibrium, the dislocation line will be semicircular. This situation can be discussed mathematically.

The same cannot be said about the next two or three stages, in fact, in the penultimate stage – which can be considered a critical step from the point of view of dislocation multiplication – we are faced with not only energetic but also crystal geometry problems. This situation is characterized by the fact that two points of the dislocation line starting from the fixed endpoints of the original dislocation segment meet each other during the growth of the area rubbed by it, namely from the opposite side compared to the dislocation line bent in the first step of the process. At the point of contact – considering the direction of the Burgers vector of the dislocation and the parallelism of the tangent drawn at this point to the dislocation line – sections of screw dislocation meet.

According to the literature, the contact screw dislocations have opposite signs at this point. If we talk about opposite sign, we can obviously first think of some vector quantity. However, the two screw dislocation segments at the contact point cannot have opposite signs since both belong to the same dislocation line. At the same time, the opposite orientation of the dislocation line vector may also arise, but we have no fixed rules regarding the orientation of this. On the other hand, in the literature – be they Hungarian, German, or English sources – the reason for the break of the dislocation line is the opposite sign of some feature or characteristic.

Without clarifying this question, it is difficult to accept the reality of the Frank–Read mechanism, which would direct our attention towards the development of new but similar mechanisms, for example when rethinking the physical phenomenon of plastic deformation and, therefore, when developing computer models of plastic deformation. Another doubt is the unanswered question of how the lower section of the loop formed at point K marked in Figure 1 can move in the opposite direction to the original dislocation section and to the direction of the sliding stress causing its buckling.

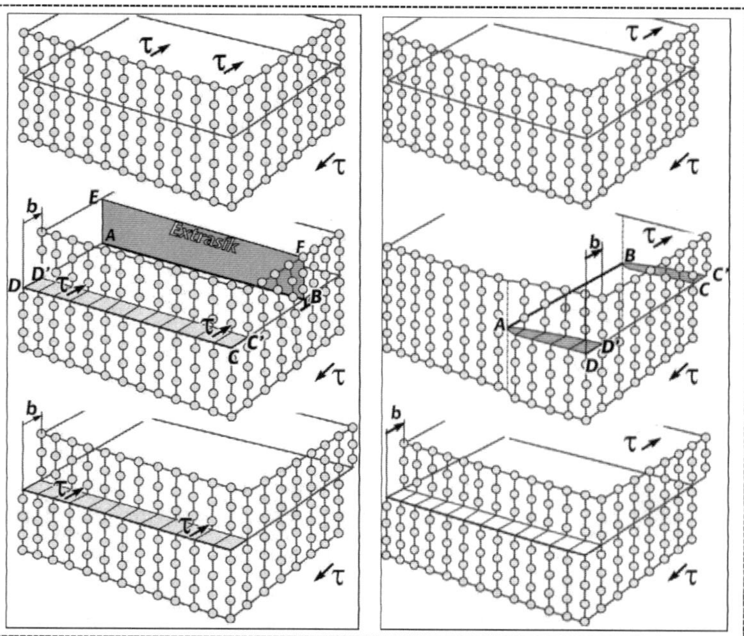

Figure 3. The usual representation of dislocation motion for edge dislocations (left) and screw dislocations (right)

Finally, it would be worthwhile to ask the question again and satisfactorily answer why we are talking about dislocation movement – both in terms of screw and edge dislocations – while it is also clearly visible on the spatial sketches depicting dislocation movement (Figure 3) that in fact it is not

IOP Conf. Series: Materials Science and Engineering **1313** (2024) 012009 doi:10.1088/1757-899X/1313/1/012009

(not only) the dislocation moves, but all the atoms of the entire space quarter above the slip plane. So, dislocation movement is a one-way movement of sub-blocks, packets, bands of crystallites – in the case of single crystals and single-crystal micropillars (Figure 4 [17]), even the entire volume above the sliding surface. However, these displacements are not consistent with the loop formation hypothesis of the Frank–Read mechanism.

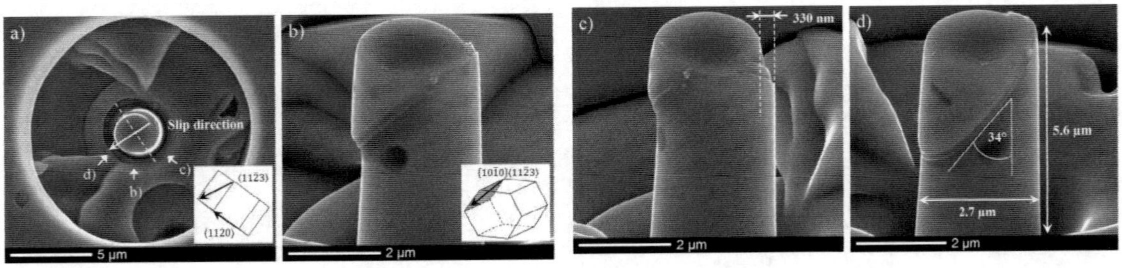

Figure 4. Typical deformation character of a single crystal micropillar in article of Dusza et al. [17]. The sliding took place in the entire upper part of the crystal.

3. Alternative mechanisms of dislocation multiplication other than the Frank–Read source

Ever since *Frank* and *Read* published [14] the dislocation multiplication mechanism named after them in 1950, this process has become an integral part of the relevant literature, and this is still true today. Since some steps of the process of the Frank–Read mechanism are difficult to interpret, and there is relatively little experimental evidence for this, alternative solutions have also been proposed during the past decades.

We list some of these based on the revolutionary book by *István Kovács* and *Lehel Zsoldos* [18].

- The further growth of dislocations already existing in the core and reaching the interface between the solid and the liquid.
- Due to the temperature gradient, the dislocation multiplication occurring in the core because of the thermal stresses and the further growth of the loops when they reach the solid / liquid interface.
- Poor quality overgrowth (epitaxy) caused by contamination.
- Multiplication originating from surface defects of the crystal because of thermal stresses.
- The impurity atoms, which are always present, mostly remain in the non-crystalline phase. Therefore, the impurity of the melt (solution) will be higher, and thus the coefficient of thermal expansion of the material crystallized in the last part of the crystallization will be different. This can lead to significant internal stress and thus dislocation multiplication.
- Formation of closed dislocation loops as a result of the condensation of vacancies in excess of thermal equilibrium. According to experience, it is mainly significant in suddenly cooled materials and in the irradiation damage process.
- The multiplication of dislocations formed in other ways as a result of thermal stresses.

In connection with the listed alternative dislocation multiplication mechanisms, it is worth looking for answers to the following questions:

- Are there dislocation multiplication mechanisms other than the Frank–Read mechanism in the literature?
- According to literature data, how can the alternative mechanisms be ranked based on their reality, and is the role of the most realistic mechanism comparable to that of the Frank–Read mechanism?
- The known alternative mechanisms are not closely related to the plastic deformation process. Is there an alternative mechanism known in the literature that works during plastic deformation? Can we think of, for example, the dressing (skin-pass rolling) of mild steel?
- Should the structure type of the crystal be considered when interpreting the Frank–Read mechanism? In the literature, crystals with a primitive, regular system are mostly associated with the movement of complete dislocations. The question is, therefore, whether the Frank–Read

mechanism occurring in the known crystal structure variants of metals and alloys can be interpreted in a similar way?
- Should the role of partial dislocations be considered, like that of cross slip?

4. Discussing the contradictions with artificial intelligence

In order to be able to discuss the emerging doubts with the most broad-minded discussion partner who can see the literature in its entirety, we invited the most well-known player in artificial intelligence, ChatGPT, to discuss this issue. The discussion took place in several rounds. It was not only useful, but the self-management process of the artificial intelligence also stood out brilliantly. In this article, we present only a short part of the entire discussion (the first nine questions and the answers), because the entire discussion and its evaluation, complete with explanations, are only in the first stage of publication on the scientific blog of *Balázs V*erő [19].

Question 1: What is dislocation multiplication according to the Frank–Read mechanism?

ChatGPT: According to the Frank–Read mechanism, dislocation multiplication is a process in which new dislocations are formed in a material through the formation and growth of a closed dislocation loop.

Question 2: What process leads to the breaking of the dislocation loop?

ChatGPT: The internal stress of the loop is greatest at the two ends where the dislocation is attached. As the loop increases, the tension increases until it exceeds the tension at the anchor points. Then the dislocations at either end of the loop begin to slip and break away from the anchor points, leaving behind a new free dislocation. In summary, dislocation loop rupture is the result of an increase in internal stress and an excess of stress at the anchoring points.

Question 3: In the literature, this is explained differently, by the intersection of dislocation segments with opposite signs. What do you think of this commonly accepted explanation?

ChatGPT: You're right. I apologize for my answer, which was not entirely accurate. The accepted explanation of the break of the dislocation loop is indeed based on the meeting of two dislocation sections with opposite signs. When the Frank–Read loop grows, the dislocation segments on its two sides approach each other. When two screw dislocations of opposite sign meet, they disappear.

Question 4: What causes the two screw dislocations to have opposite signs at the meeting point? Along the line, the Burgers vector is constant, so only some other characteristic can cause the opposite sign. What other feature of the dislocation line can cause the opposite sign?

ChatGPT: You are right, the Burgers vector is constant along a given dislocation. /…/ In the case of screw dislocations, the angle between the dislocation line and the Burgers vector is the critical characteristic: this angle is 90°. The opposite sign does not result from the direction or magnitude of the Burgers vector, but from the direction of the twist around the dislocation line. When two screw dislocations meet and their direction of twisting is opposite (one shows a clockwise twist and the other counterclockwise), then these dislocations are considered to be of opposite sign. In this case, if the two dislocations meet, the twists neutralize each other. In summary, the opposite sign of twist dislocations is caused by the direction of the twist, not the direction of the Burgers vector or the dislocation line.

Question 5: Answer as a qualified, experienced material scientist with modern knowledge!

ChatGPT: Of course, I will continue to answer your questions as a materials scientist.

Question 6: Some steps of the Frank–Read mechanism are difficult to interpret, there is little experimental evidence, and alternative solutions have also been created. Do you know more?

ChatGPT: Yes. The ones you listed cover many cases, but there are also specific mechanisms: 1) Dislocations arising during phase transformations, e.g. the martensitic transformation. 2) Some alloys or impurities can create dislocation nuclei. 3) Ion beam irradiation. 4) Crystal growth defects, e.g. nucleation defects, growth spirals. 5) Plastic deformation.

Question 7: In addition to the Frank–Read dislocation multiplication mechanism, is there a known mechanism that works exclusively during plastic deformation?

ChatGPT: Yes, in addition to the Frank–Read mechanism, there are also those that operate exclusively during plastic deformation. Such is the Taylor–Orowan mechanism. This occurs especially in materials where the movement of dislocations is significantly hindered, such as fine-grained materials or those in which dislocations interact with precipitates or impurities.

Question 8: Should the Frenkel or Peierls model be taken into account when the loop is formed?

ChatGPT: When the dislocation loop is formed during the Frank–Read mechanism, the movement of the dislocation in the crystal lattice is interpreted based on the Peierls model. The Frenkel model describes the formation and movement of dislocations in ionic crystals, where Coulomb interactions dominate. In the context of the Frank–Read mechanism, the Peierls stress is the minimum stress required for a dislocation to cross the Peierls barrier and move in the crystal lattice.

Question 9: Can this answer be interpreted as the Peierls force being able to overcome the effect of the shear stress on the dislocation line during the growth of the loop?

ChatGPT: Your question touches on a very interesting and important aspect of dislocation motion.

In the Frank–Read mechanism, as the dislocation loop grows, the shear stress from the external loading must be sufficient to overcome the Peierls force. If the shear stress is insufficient, the dislocation cannot move, and the loop will not grow further. So, my answer to your question is yes, as the dislocation loop grows, the shear stress from the external load should be able to overcome the Peierls force.

We repeat that the entire discussion on the topic can be found on the material science.eu blog. By the way, the conversation was held in Hungarian, but the translation into any language can also be easily implemented today, since one of the most spectacular areas of application of artificial intelligence is the transition between languages.

5. Forgotten or almost unknown approaches to dislocation propagation mechanisms

In this chapter, we list some of the opinions regarding dislocation multiplication that did not join the enormously growing chorus that would consider the Frank–Read mechanism valid without any criticism for dislocation multiplication.

The Peach–Koehler analysis is particularly important from the point of view of the operation of the Frank–Read source, which deals in detail with the determination of the "critical stress" necessary for the operation of a dislocation fixed at two points as a source [20].

Kovács and *Zsoldos* note that the contradictions of the Frank–Read theory have been brought up by many, e.g. *Johnston* and *Gilman* [21] arguing that the speed of dislocation movement is too low, and *Tetelman* arguing the primacy of dipole formation [22]. Accepting *Gilman*'s criticism, in 1973 *Reed-Hill* wrote that experimental evidence of dislocation nucleation indicates that Frank–Read sources do exist in crystals, but it is not known how important these dislocation generators are in the ductile deformation of metals. They also add that the latest evidence (at the time) shows that dislocations can form without the help of Frank–Read or similar sources, and that if the dislocations are not formed by dislocation generators, they must be formed by nucleation processes [23].

Smallmann says that the simplest type of source that is due to *Frank* and *Read*, and accounts for the regenerative multiplication of dislocations. A modified form of the Frank–Read source is the multiple cross glide source, first proposed by *Köhler*, which, as the name implies, depends on the cross-slip of screw dislocations and is therefore more common in metals of intermediate and high stacking-fault energy [24].

In the book published in 1998 by *Meyers* and *Chawla*, the evaluation is formulated, according to which it can be clearly stated that Frank–Read sources have only been observed in metals in a few cases. Contrary to our experience, they believe that the original formulation of the Frank–Read source is not widespread, but only its various modifications – the node formation and the cross-slip – may be important mechanisms of dislocation generation after the first few percent of plastic deformation [25].

Estrin, Kim and *Nabarro* investigated the dislocation generation process in nanocrystalline metals using the Frank–Read mechanism as a dislocation source. They concluded that the functioning of this mechanism – according to the principle that declares the dependence of the strain rate sensitivity (SRS) on grain size – is only possible in a very limited range of the source length and stress. The calculated and the experimental data for copper demonstrated clearly that the Frank–Read mechanism can be excluded [26].

Golosova et al performed numerical simulation to observe the formation of dislocation loops. It was found that the resulting minimum stress of the dislocation source coincides only in a rather narrow range of the length of the sources, according to the classical theory, the so-called with critical Frank-

Read stress. Beyond the front of dislocations spreading further from the dislocation source, only the dislocation loops formed by the Orowan mechanism remain, the length of which is 12–14 % of the length of the originally formed loop [27].

As we saw above that a research result could be called a dramatic proof or a decisive step, we believe that it is no exaggeration to call *Murr*'s article a huge and brave taboo decision. Already in the title of his article, *Murr* goes against the decades-old canon, since he calls the Frank–Read mechanism a myth [28]. *Murr* writes that *J.C.M. Li* already questioned it in the early 1960s the dislocation pileup phenomenon at grain boundaries and proposed so-called grain boundary ledges as sources for dislocations, in contrast to Frank–Read sources. In his article, *Murr* reviews the proposals and evidence in question based on transmission electron microscopy of stainless steel and nickel. The test results support his conclusion that Frank–Read sources are practically absent from the plastic deformation process of metals and alloys, but the ledges on the interfaces dominate as dislocation sources.

Conclusions

Above, we presented all the considerations that we could gather about the theory of dislocation multiplication based on the Frank–Read sources. However, there is an even more fundamental contradiction, which we consider necessary to mention, to recall precisely, since *Kovács* and *Zsoldos* [18] drew attention to this contradiction decades ago [20]. It is the St. Venant compatibility condition. In order to establish the general continuum theory of lattice defects, the equations of classical elasticity must also be generalized, because lattice defects are internal sources of elastic deformations that cannot be described by a single-valued and continuous displacement vector everywhere throughout the body, so these deformations do not satisfy the St. Venant compatibility condition are fulfilled.

Put somewhat more simply, the contradiction is that the stress fields around the dislocation cannot be explained by the effect of the external stress, yet we claim that the dislocations arise from dislocation sources activated by the external stress field. Encouraging results have already been achieved in this direction in recent years, such as Lubarda's work [29].

To this we could also add simple considerations such as geometric features that can be calculated based on the largest values of the lattice parameter and dislocation density, or surprisingly small values calculated for the speed of dislocation movement.

We believe that investigating the plastic deformation of micropillars that can be created with FIB is a particularly useful tool in the hands of researchers, as metals and alloys used in industrial practice can be studied with the help of this tool; n excellent example of this is the work of *Dimiduk* and his colleagues [30].

However, all these considerations should be thought through by a new generation of researchers, whose members know how to effectively manage today's research infrastructure and are able to examine with a critical eye the theorems that are believed to be eternally valid.

Acknowledgements

The authors express their gratitude to the University of Dunaújváros and Ecotech Nonprofit Zrt. for the promotion of the research results, which were supported by the "National Research Development and Innovation Fund" of the Ministry of Innovation and Technology, the Topic Excellence Program 2020 (2020-4.1. 1-TKP2020) were financed by the tender program.

References

[1] Rejtő S 1907 A maradó alakváltozásokra vonatkozó kísérleti eredmények a brüsszeli kongresszuson. *A Magyar Mérnök és Építész Egylet közlönye* **41** 66-79 [in Hungarian]
[2] Rejtő S 1914 A maradó, vagyis az ideigtartó és a maradandó alakváltozások mechanikájának alaptételei és alkalmazásuk. *Mathematikai és természettudományi értesítő.* **32** 257-317
[3] Rejtő S 1925 A maradó alakváltozások mechanikájának alkalmazása a textiliparban. *Mathematikai és természettudományi értesítő.* **41** 1-8
[4] Misángyi V 1908 A maradó alakváltozások. *A Magyar Mérnök és Építész Egylet közlönye.* **42.** 70-98

[5] Gállik I 1909 A szakadási kontrakció jelentősége a vasanyagok minőségének megítélésében, különös tekintettel a képlékenységre. *A Magyar Mérnök és Építész Egylet közlönye*. **43** 217-225
[6] Verő J 1931 Nemvasfémek képlékeny alakításának viszonyai. *Bányászati és kohászati lapok*. **64** 355-359
[7] Geleji S 1942 A képlékeny alakítás néhány elméleti problémája. *Bányászati és kohászati lapok*. **75** 316-319
[8] Kármán Th. von 1910 *Untersuchungen über die Bedingungen des Bruches und der plastischen Deformation, insbesondere bei quasi-isotropen Körpern. Habilitationschrift* (Göttingen: Universität Göttingen)
[9] Nádai A 1927 *Der bildsame Zustand der Werkstoffe* (Berlin: Julius Springer)
[10] Orowan E 1934 Zur Kristallplastizität. III. Über den Mechanismus des Gleitvorganges. *Zeitschrift für Physik*. **89** 634–659
[11] Polanyi M 1934 Über eine Art Gitterstörung, die einen Kristall plastisch machen könnte, *Zeitschrift für Physik*. **89** 660–664
[12] Taylor G I 1934 The mechanism of plastic deformation of crystals. Part I. Theoretical. *Proceedings of the Royal Society of London*. **145** 362–387
[13] Paterson N S 2013 *Materials Science for Structural Geology* (Dordrecht: Springer Science)
[14] Frank F C and Read W T 1950 Multiplication Processes for Slow Moving Dislocations. *Phys. Rev*. **79** 722–723
[15] Dash W C 1959 Growth of Silicon Crystals Free from Dislocations. *J. Appl. Phys*. 30, 459–474
[16] Philibert J, Vignes A, Bréchet Y and Combrade P 1998 *Métallurgie du minerai au matériaux* (Paris: Masson)
[17] Dusza J, Csanadi T, Nemeth D, Grasso S and Reece M 2015 Nanoindentation, micropillar compression and nanoscratch testing of ZrB2 grains. in: Legros M (ed) *Nanomechanical Testing in Materials Research and Development V*, paper **82**
[18] Kovács I and Zsoldos L 1973 *Dislocations and plastic deformation* (Oxford: Pergamon Press)
[19] Verő B 2023 https://www.anyagtudomany.eu
[20] Peach M and Koehler J S 1950 The Forces Exerted on Dislocations and the Stress Fields Produced by Them. *Phys. Rev*. **80** 436 –439
[21] Johnston W G and Gilman J J 1959 Dislocation Velocities, Dislocation Densities, and Plastic Flow in Lithium Fluoride Crystals. *J. Appl. Phys*. **30** 129–144
[22] Tetelman A S 1962 Dislocation dipole formation in deformed crystals. *Acta Metallurgica*. **10** 813–820
[23] Reed-Hill R E 1973 *Physical Metallurgy Principles* (Boston: PWS Publishing Co.)
[24] Smallmann R E 1985 *Modern Physical Metallurgy, 4th edition* (Butterworth-Heineman)
[25] Meyers M A and Chawla K K 1998 *Mechanical Behaviors of Materials* (Pearson Prentice Hall)
[26] Estrin Y, Kim H S and Nabarro F R N 2007 A comment on the role of Frank–Read sources in plasticity of nanomaterials. *Acta Materialia*. **55** 6401–6407
[27] Golosova T N, Slobodskoi M I and Popov L E 1992 Modeling of a dislocation source in the field of activated and unactivated discrete barriers. *Russian Physics Journal*. **35** 912–916
[28] Murr L E 2016 Dislocation Ledge Sources: Dispelling the Myth of Frank–Read Source Importance. *Metall Mater Trans A*. **47** 5811–5826
[29] Lubarda V A 2019 Dislocation Burgers vector and the Peach–Koehler force: a review. *J Mater Res Technol* **8** 1550
[30] Dimiduk D M, Uchic M D and Parthasarathy T A 2005 Size-affected single-slip behavior of pure nickel microcrystals. *Acta Mater* **53** 4065–4077

IOP Conf. Series: Materials Science and Engineering **1313** (2024) 012010 doi:10.1088/1757-899X/1313/1/012010

Physical Simulations of the Tool Degradation in Friction Stir Welding

Nour El Imane Djimaoui [1], Yoni Adonyi[2], Jeroen De Backer[2,3], Árpád Kovács[1], László Kuzsella[4], Valéria Mertinger[1]

[1]Institute of Physical Metallurgy, Metal-forming and Nanotechnology, University of Miskolc, Hungary

[2]Welding Technology Center, University West, Sweden

[3]Robotic Friction Stir Welding Group, TWI Ltd, Sheffield Labs, UK

[4] Institute of Materials Science and Technology, University of Miskolc, Hungary

E-mail: femvali@uni-miskolc.hu

Abstract The present investigation is part of a comprehensive failure study of friction stir welding (FSW) tools, where experimental non-destructive and destructive evaluations are combined with analytical- and physical- simulations of tool wear mechanisms. This paper focuses on the research methodology and the primary results of Gleeble simulations of thermal- and mechanical-cycle effects on an H13 steel material widely used for FSW tools for welding aluminum and magnesium alloys.

1. Introduction

FSW Friction Stir Welding is a modern solid-state joining technique which has gained significant attention in the manufacturing industry due to its numerous advantages, such as the ability to weld dissimilar metals, low residual stresses and distortions, and no need for filler wire. Developed by The Welding Institute (TWI) in 1991, FSW has been used for high volume production since 1996 in many industries from automotive to aerospace and nuclear. It is a joining process that uses a non-consumable tool to join two facing workpieces without melting the workpiece material [1].FSW employs various tool materials, such as hot work tool steels (for example, H13), high-speed steels, superalloys (like MP159), and cemented carbides. However, one of the challenges of FSW is the gradual degradation of the tool material and wear after tens of meters of weld made (depending on the material) [2]. Tool wear impacts not only the quality of produced welds, the mechanical properties, and its variability along with the welding cycles but also affects the dimensional features of the tool, reducing the tool life cycle and thus increasing production costs and reducing productivity [3]. This issue became a more acute problem when steel, titanium, copper, and superalloys were FSW welded, primarily due to the much higher temperatures experienced (up to 1200-1300°C as opposed to 450-550°C in aluminum) and the much higher flow stress of these base metals. At the same time, the H13 tool used for aluminum did not perform well when thick aluminum sections of up to 50 mm were welded in the early 2000s.

Various studies have explored the degradation mechanism of FSW tools. One such study was conducted by T. Majeed et al., who analyzed the tool degradation of high thickness ratio unequal thickness dissimilar material joints made between 6.3 mm thick AA2024-T3 and 2.5 mm thick AA7475-T7 plates; they found that the pin undergoes significant degradation in the form of wear and deformation compared to the tool shoulder, and lower flow stresses caused by higher process temperature lead to lower tool wear and deformation [4]. In a study by S. Hanke et al., the degradation of FSW tools was investigated during the FSW process of nickel-base alloy 625, using polycrystalline boron (PCBN) tool material. The study found that during FSW of Ni-base alloy 625 using PCBN tools, the wear mechanisms include abrasive and adhesive wear. The thermal softening of the tungsten-rhenium binder phase and dissolution of tungsten into

Inconel 625 are potential mechanisms for early wear and failure of the PCBN tools for FSW of Inconel 625[5]. P. Sahlot investigated the quantitative wear analysis of H13 steel tool during friction stir welding of Cu-0.8%Cr-0.1%Zr alloy and observed a higher total tool wear for faster tool rotational speeds [6]. The study conducted by A. Berrais et al. sheds light on the wear behavior of nitride AISI H13 hot work tool steel during an aluminum hot extrusion process. The study revealed that the primary wear mechanisms involved sticking/adhesion, cracking, oxidation, and detachment of material [7].

El-Shenawy and J. Marashi conducted a study using the Gleeble physical simulator to explore tool degradation and wear during thermal-mechanical processing, and their research concluded that the simulator's ability to control and simulate thermal cycles makes it an ideal tool for investigating hot ductility and assessing the deformation of tool materials, such as H13 tool steel, under hot forging conditions [8], [9]. The Gleeble equipment couples the ability to apply thermal cycles and mechanical deformation to material specimens. Direct resistance heating is applied to the test specimen according to a prescribed time-temperature cycle input with the system's workstation computer. The ability to program a desired thermal cycle allows users to simulate such processes as welding. Concurrent with thermal processing, specimens can be deformed in tension or compression [10].

We typically have no information about the external influences during the FSW process, such as temperature and wear, and therefore, we do not know the exact mechanism of tool failure. The study aimed to perform a comparative analysis of a tool damaged during welding and a sample condition trained in a physical simulator (Gleeble 3800) of the same alloy to better understand the real tool failure processes.

2. Materials and methodology

TWI Ltd provided the H 13 FSW tool where the failure occurred after just one stage of welding, which was unexpected as the tool was supposed to have a long lifespan. For the Gleeble simulation, the test specimens were made from the same H13 alloy. The geometry of the FSW tool and the Gleeble specimen is shown in Figure 1.The simulation sample was made into a cylindrical piece with ϕ10 mm × 100 mm dimensions. Before the Gleeble simulation, the usual hardening heat treatment of the steel was carried out. Austenitetization at 982°C followed by oil quenching and tempering at 150°C. The main alloying elements (except carbon) were checked for both samples. The results of energy dispersive microprobe (EDX) analysis are shown in Table 1.

Table 1. Concentration of main alloying elements of FSW tool and Gleeble samples

	Main alloying elements, wt%					
	Si	**Mo**	**V**	**Cr**	**Mn**	**Fe**
FSW tool	1.43	1.86	1.21	5.41	0.64	89.44
Gleeble sample	1.17	1.58	1.15	5.30	0.60	90.2

In the friction stir welding of aluminium at TWI Ltd, the actual recorded tool pin temperature variation is depicted in Figure 2 which fluctuates between 400-600°C. On the other hand, Figure 3 shows the interpretation of the forces the tool experienced during FSW, with the maximum load being 8 KN.

To determine the significance of thermal and mechanical effects experienced by the tool during the FSW process of aluminum, three types of physical simulation cycles were applied to the Gleeble samples. These cycles are designed to replicate the conditions encountered during the process. While thermal effects arise from the generated heat, mechanical effects arise from the forces applied to the tool during rotation and axial forces. Simulating these conditions allows us to determine which result holds more significance. Accordingly, purely thermal, purely mechanical and combined thermal and mechanical simulations were performed.

During the first simulation, the sample was subjected only to thermal cycles. It was heated to 600°C at a rate of 300°C/s for 1 second, then cooled to 400°C at a rate of 100°C/s for 1 second, and this process was repeated for a total of 15 cycles, as shown in Figure 4-a. The mechanical simulation involved subjecting the H13 steel sample to 15 cycles of force variation between 0-8 KN at a temperature of 600°C. The simulation assessed the material's behaviour under cyclic loading and a constant temperature (see Figure 4-b). In the last case, the sample was subjected to thermal and mechanical simulation, as depicted in Figure 4-c.The piece was heated to a temperature of 600°C and subjected to a tensile loading of 8 KN for 5 minutes, both of which were done in a steady-state manner.

Three parallel samples were subjected to each simulation. At the end of the simulation, one sample was fractured at room temperature during the tensile test, and one was metallographically analyzed in longitudinal and cross-sectional dimensions. The fractured surface was also examined in microscopy. The tensile test and microscopic examination were done using the Instron universal test machine and Zeiss EVOMA 110 scanning electron microscope (SEM). The FSW tool's work surface and cross-section at the bottom part were also on the SEM test.

FSW tool Gleeble specimen

Figure 1.FSW tool and the Gleeble sample.

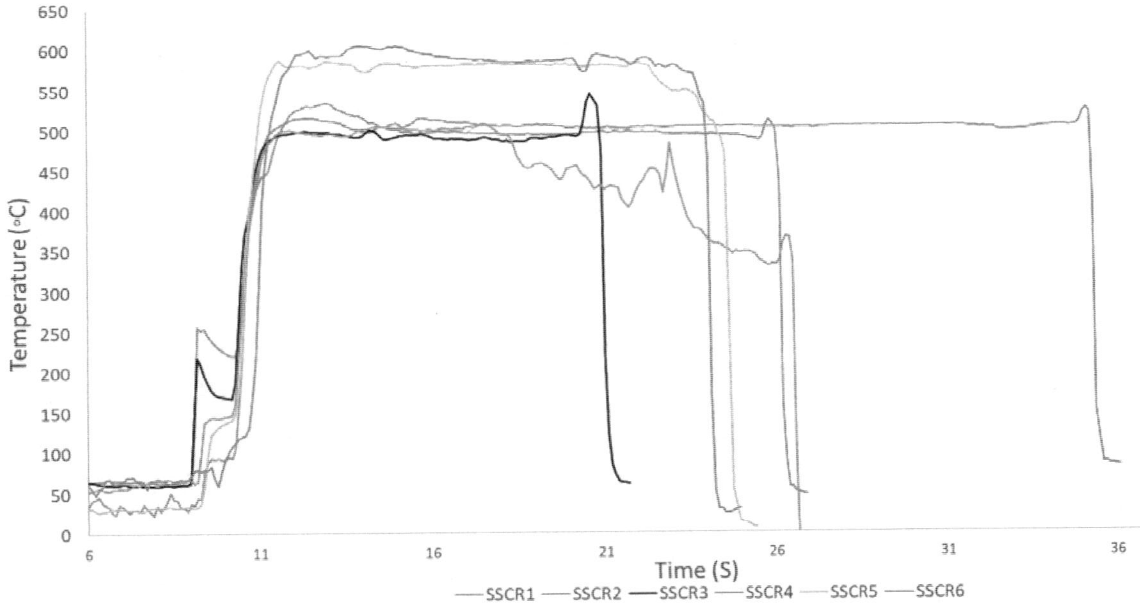

Figure 2.Temperature variation in the tool pin during FSW of aluminium.

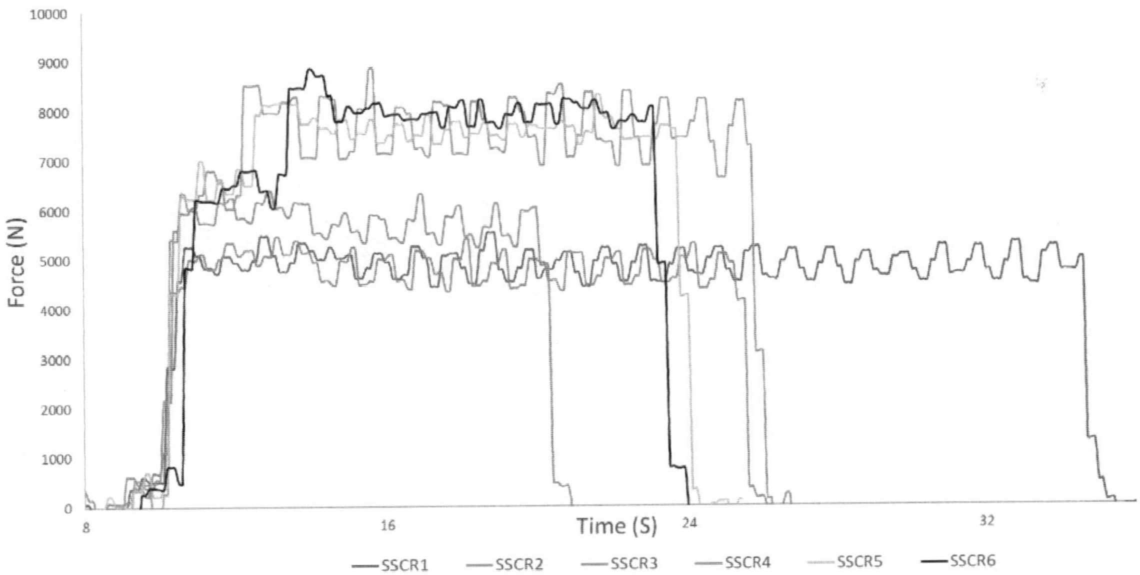

Figure 3.Force variations in the tool pin during FSW of aluminium

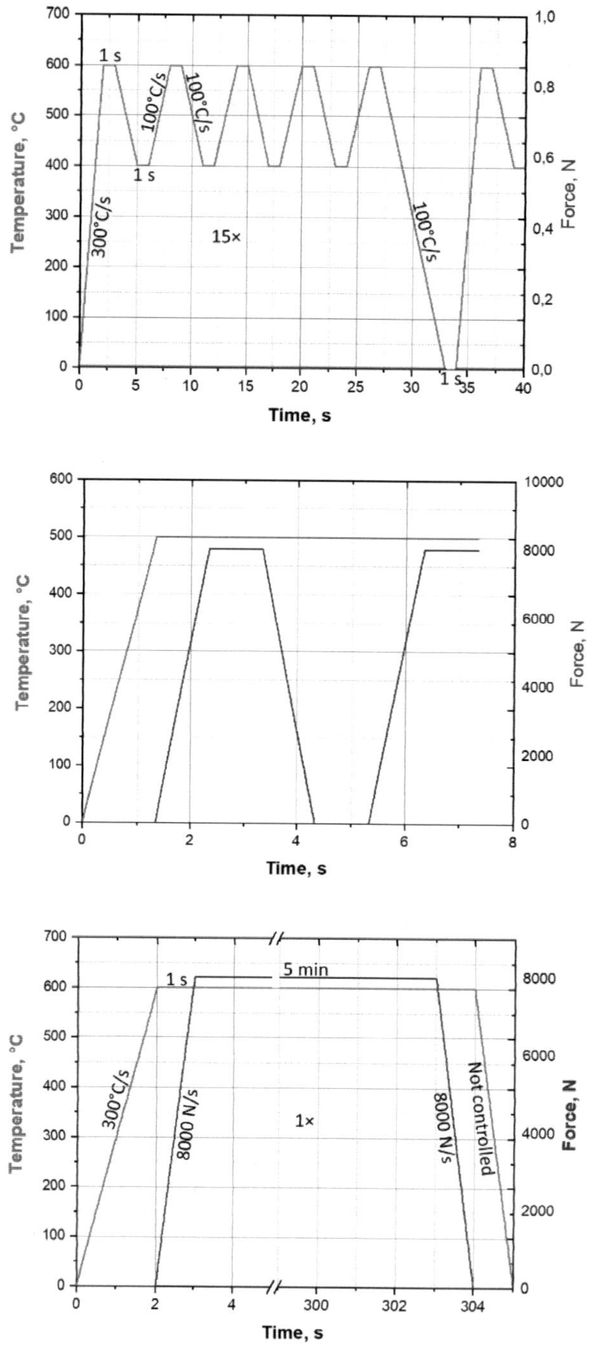

a)

b)

c)

Figure 4.Simulation cycles: a) thermal only b) mechanical only c) thermal + mechanical

3. Results

Figure 5 compares the SEM image of the bottom part of the H13 tool, which is not affected by the weld process, with the initial state of the gleeble specimen. The comparison shows that the specimens are in the same condition in terms of both grain size and martensite structure. This finding was confirmed by hardness testing. HV03= 500 was measured on both specimens.

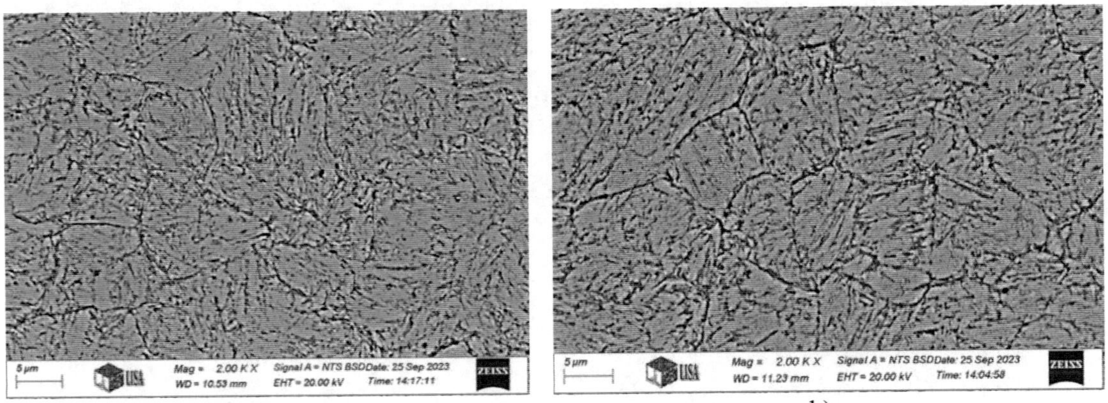

a) b)

Figure 5.SEM images a) bottom part of H13 FSW tool, b) H13 steel microstructure of Gleeble samples in the initial condition

After examining the working surface of the FSW tool, it was found that cracks and adhesive wear were present, as shown in Figure 6. The appearance of cracks is completely unjustified at this early stage in the tool's life.

The tensile test results of the Gleeble sample after cyclings are shown in Table 2. The data demonstrates that thermal and mechanical simulation significantly impacts the sample material. It also shows that the thermal simulation has a more significant impact on the material than the mechanical simulation, which has the weakest impact on the sample material.

The microstructures of initial and various simulated samples are illustrated in Figures 7. The simulated samples have a microstructure similar to the initial sample microstructure, which is martensitic, with a magnification of 2.00 KX. Figure 8 shows a higher magnification of 7.50 KX, which reveals the presence of Vanadium carbides in the samples that were simulated thermally only and thermally and mechanically. The existence of Vanadium carbides can explain the difference in maximum tensile stress between each sample.

Table 2. Tensile stress of each simulated sample after Gleeble cycling

Sample condition	**Initial**	**Thermal only**	**Mechanical only**	**Thermal +mechanical**
Tensile stress, MPa	2061	1805	1868	1748

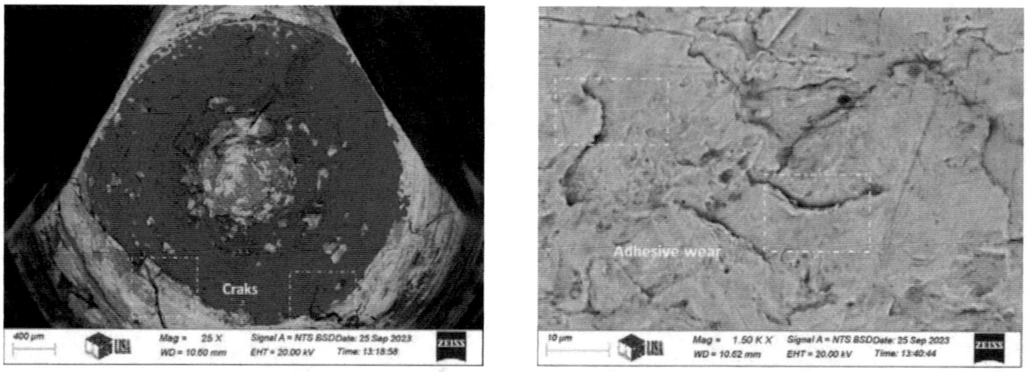

Figure 6. Failure in the surface of the H13 FSW tool

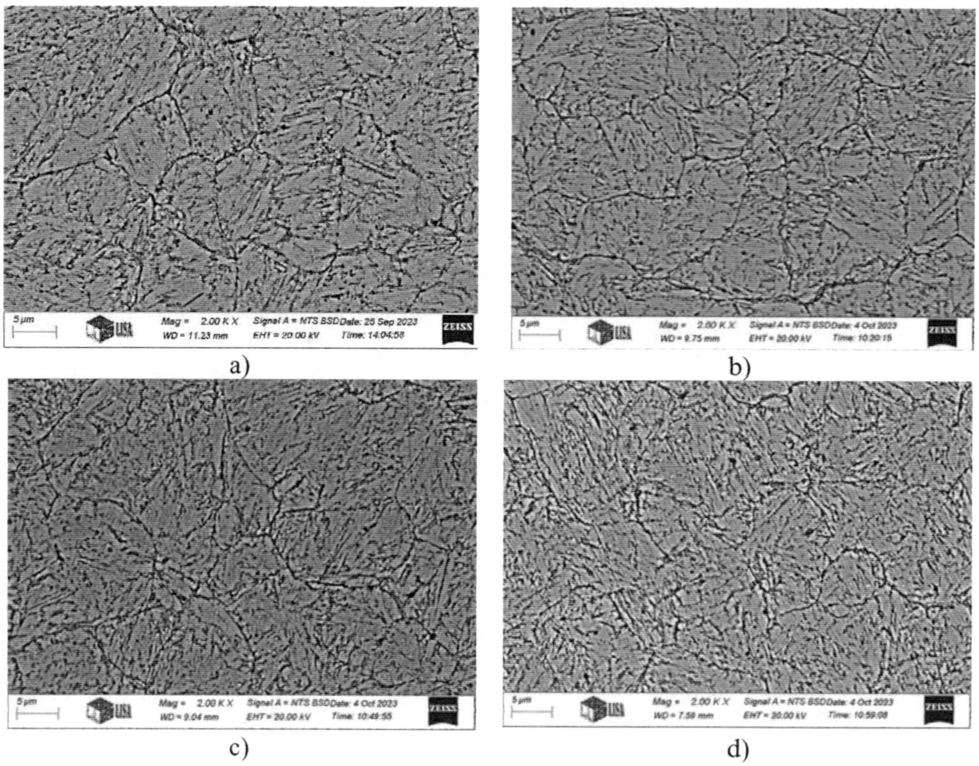

Figure 7.Microstructure of Gleeble sample: a) initial/unsimulated, b) thermal-only simulation, c) mechanical-only simulation, c) thermomechanical simulation

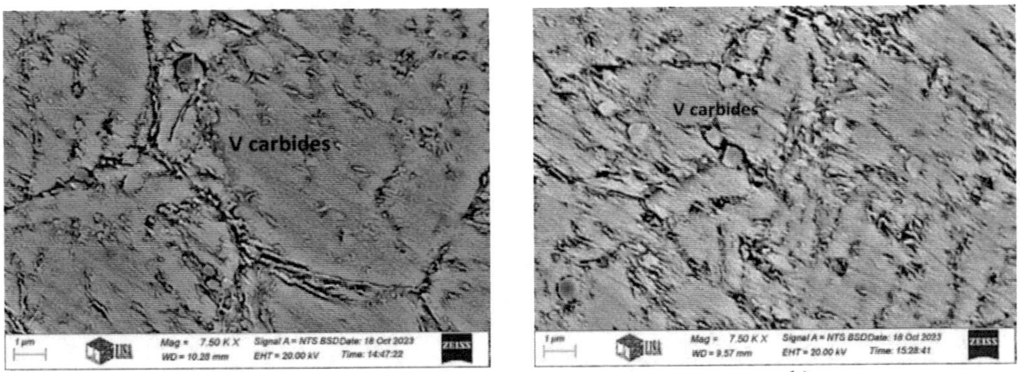

a) b)

Figure 8. Vanadium carbide was presented in the simulated samples by a) thermal-only and b) thermal and mechanical simulations

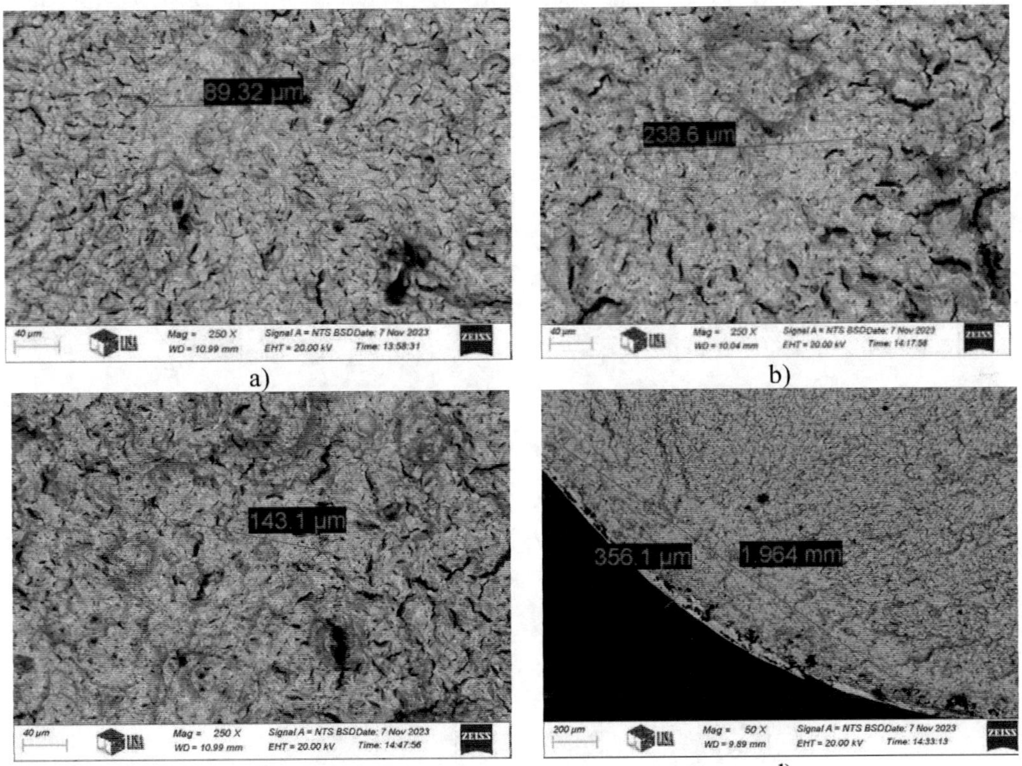

a) b) c) d)

Figure 9. Fracture surface after tensile test of the samples: a) initial/unstimulated) thermal-only simulation, c) mechanical-only simulation, and d) thermal and mechanical simulation.

IOP Conf. Series: Materials Science and Engineering **1313** (2024) 012010 doi:10.1088/1757-899X/1313/1/012010

Figure 9 shows the fracture surface of each simulated sample after a tensile test at room temperature. The diameter of the plastic deformed region with the ductile fracture is displayed. The thermal-only simulated sample and the thermal and mechanical sample show a larger diameter compared the intial one, which can be explained by the effect of the thermal factor. On the other hand, the mechanical-only sample shows a minimal difference compared to the fracture of the initial sample. These findings confirm the results of the tensile stress presented in Table 2. Figure 10 presents the fracture surface of the initial, unsimulated H13 material sample after a tensile test at room temperature. The fracture of this material exhibits a combination of ductile and brittle fractures, with two types of fractures: intercrystalline and transcrystalline fractures.

After conducting a tensile test at room temperature for each simulated sample, the resulting fractography shows the following observations:

For the sample subjected to thermal simulation only: The fractography of this sample, as presented in Figure11, reveals the presence of long cracks on the fracture surface, characterized by a combination of ductile and brittle fractures, including intercrystalline and trans crystalline fractures. The fracture surface of the sample that underwent mechanical simulation, as shown in Figure 12, did not differ significantly from the initial sample fracture (Figure 10). Similar to the initial sample fracture, the fracture types observed combined ductile and brittle fractures.Figure 13 presents the fracture of a sample simulated by thermal and mechanical methods. The dominant fracture type in this sample is an intercrystalline brittle fracture with some trans crystalline brittle fracture.

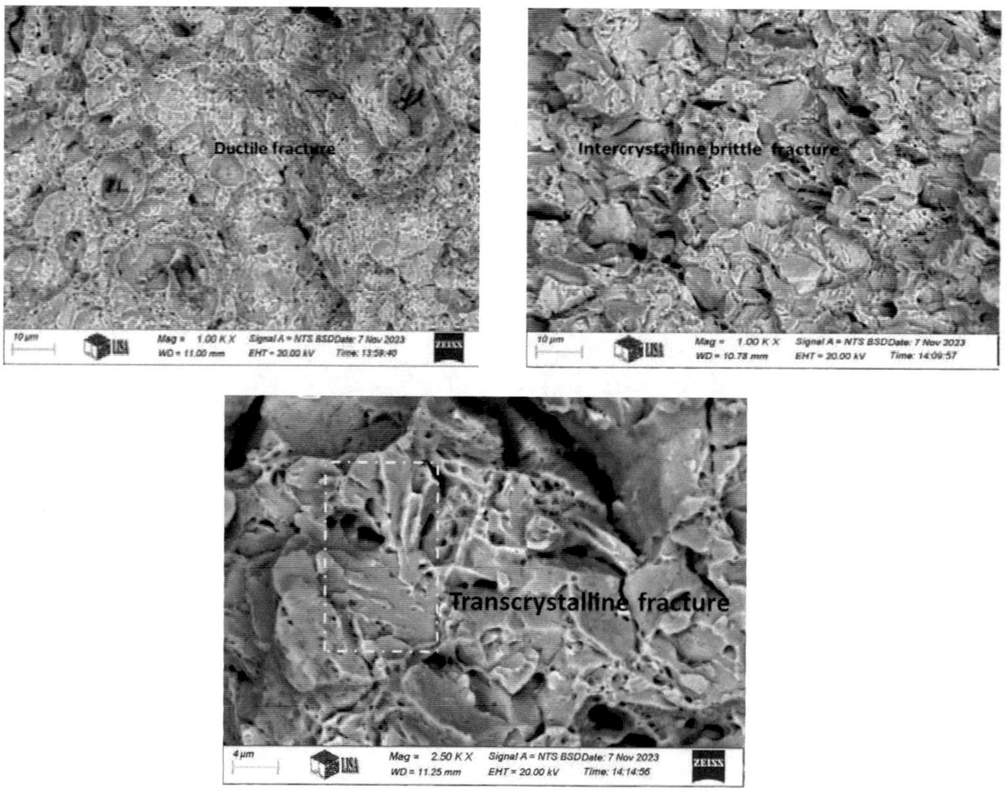

Figure 10. Different fracture types are presented in the initial sample

IOP Conf. Series: Materials Science and Engineering **1313** (2024) 012010 doi:10.1088/1757-899X/1313/1/012010

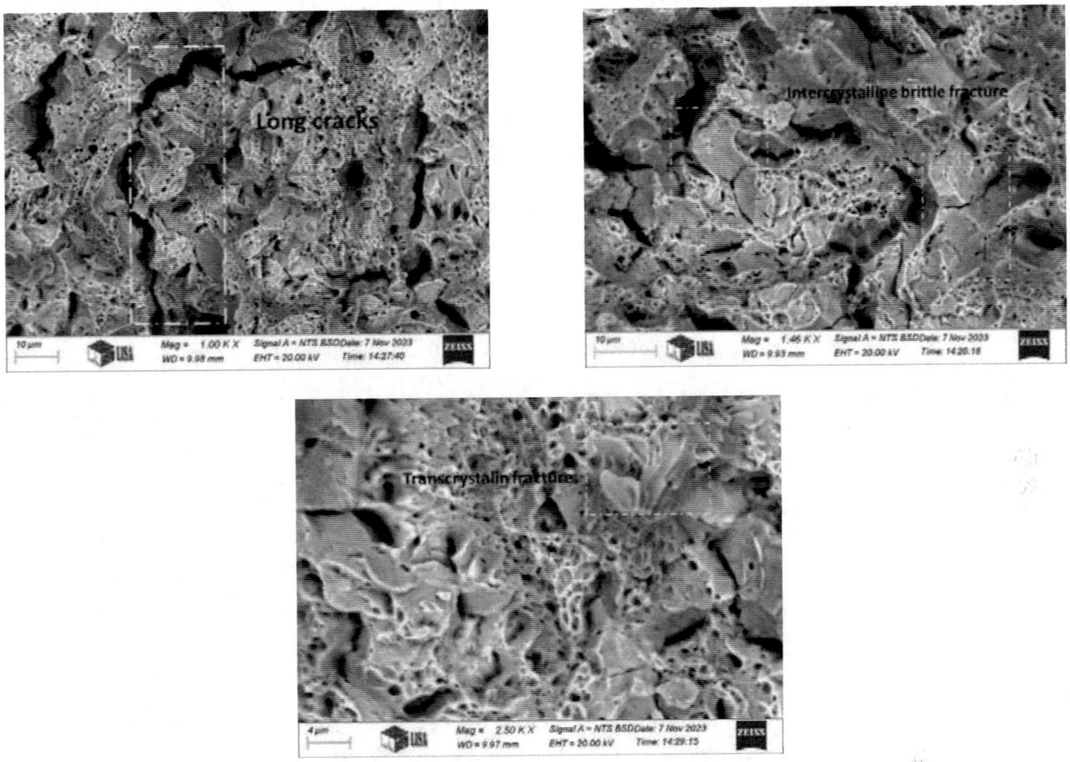

Figure 11. Fracture types are presented in the simulated sample through a thermal-only simulation

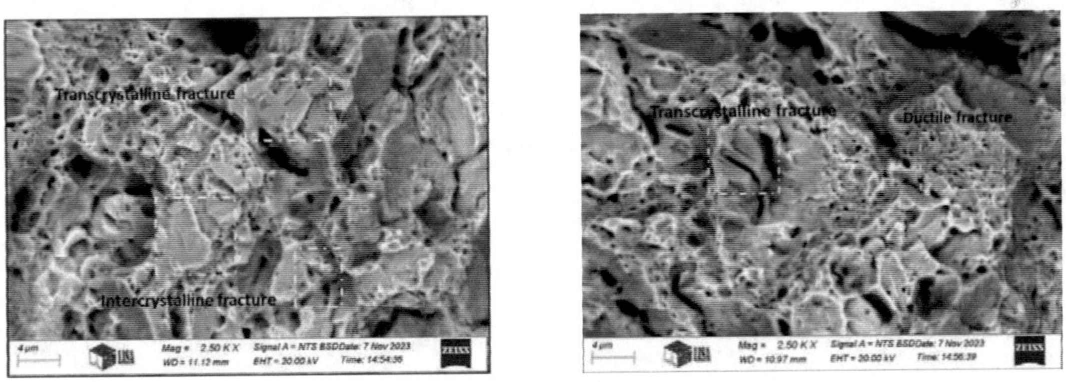

Figure 12. Fracture types are presented in the simulated sample through a mechanical-only simulation

Conclusion

We established good correlations between the tool material and simulation H13 properties and developed adequate simulation parameters that properly reflected actual FSW welding conditions. This allowed us to understand the separate effect of thermal and mechanical loading.

As part of our future work, we plan to conduct simulations at higher temperatures and increased loads, using compressive loading instead of tension and a more significant number of cycles. This strategic approach explores how intensified thermal and mechanical conditions will affect the simulated samples. We expect to gain a more nuanced understanding of material behavior, tool degradation, and wear patterns by pushing the temperature, load, and cycle parameters to their limits. These expanded simulations will provide more comprehensive insights into the response of tool materials, particularly H13 tool steel, under more extreme welding conditions.

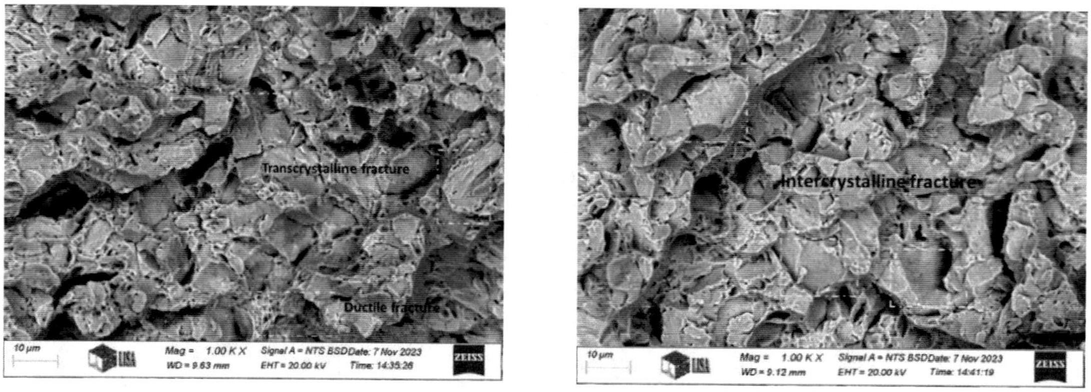

Figure 13.Fracture types are presented in the simulated sample through a thermal and mechanical-only simulation.

References

[1] X. Jiang, B. P. Wynne, and J. Martin, "Variant selection in stationary shoulder friction stir welded Ti-6Al-4V alloy," *Journal of Materials Science & Technology*, vol. 34, no. 1, pp. 198–208, Jan. 2018, doi: 10.1016/j.jmst.2017.11.024.

[2] K. Chiteka, "Friction Stir Welding/Processing Tool Materials and Selection," *International Journal of Engineering Research*, vol. 2, no. 11, 2013.

[3] C. C. De Castro *et al.*, "Tool wear mechanisms and effects on refill friction stir spot welding of AA2198-T8 sheets," *Journal of Materials Research and Technology*, vol. 20, pp. 857–866, Sep. 2022, doi: 10.1016/j.jmrt.2022.07.092.

[4] T. Majeed, Y. Mehta, and A. N. Siddiquee, "Analysis of tool wear and deformation in friction stir welding of unequal thickness dissimilar Al alloys," *Proceedings of the Institution of Mechanical Engineers, Part L: Journal of Materials: Design and Applications*, vol. 235, no. 3, pp. 501–512, Mar. 2021, doi: 10.1177/1464420720971769.

[5] "Degradation mechanisms of pcBN tool material during Friction Stir Welding of Ni-base alloy 625 - ScienceDirect." Accessed: Nov. 26, 2023. [Online]. Available: https://www.sciencedirect.com/science/article/pii/S0043164817301898

[6] P. Sahlot, K. Jha, G. K. Dey, and A. Arora, "Quantitative wear analysis of H13 steel tool during friction stir welding of Cu-0.8%Cr-0.1%Zr alloy," *Wear*, vol. 378–379, pp. 82–89, May 2017, doi: 10.1016/j.wear.2017.02.009.

[7] A. Berrais *et al.*, "Analysis of wear of a nitrided AISI H13 hot work tool steel in an aluminium hot extrusion process," *Wear*, vol. 514–515, p. 204587, Feb. 2023, doi: 10.1016/j.wear.2022.204587.

[8] E. H. El-Shenawy, "Physical simulation technology for thermo-mechanical processing of metallic alloys using Gleeble system," *Materials Today: Proceedings*, vol. 28, pp. 998–1004, Jan. 2020, doi: 10.1016/j.matpr.2019.12.339.

[9] J. Marashi, E. Yakushina, P. Xirouchakis, R. Zante, and J. Foster, "An evaluation of H13 tool steel deformation in hot forging conditions," *Journal of Materials Processing Technology*, vol. 246, pp. 276–284, Aug. 2017, doi: 10.1016/j.jmatprotec.2017.03.026.

[10] E. H. El-Shenawy, "Physical simulation technology for thermo-mechanical processing of metallic alloys using Gleeble system," *Materials Today: Proceedings*, vol. 28, pp. 998–1004, Jan. 2020, doi: 10.1016/j.matpr.2019.12.339.

IOP Conf. Series: Materials Science and Engineering **1313** (2024) 012011 doi:10.1088/1757-899X/1313/1/012011

Characterization of friction between metal-polymer material pairs with different speeds and surface pressures

R. Kohlhéb [12], R Horváth [1*], M. Réger [1]

[1]Óbuda University, Donát Bánki Faculty of Mechanical and Safety Engineering, H1081, Népszínház u. 8., Hungary
[2]Doctoral School on Materials Sciences and Technologies, Óbuda University, Budapest
*horvath.richard@uni-obuda.hu

Abstract. Engineering design requires accurate knowledge of the friction conditions of material pairs. The friction coefficient (static or dynamic) depends on many parameters in addition to the pairs of materials used (e.g. surface roughness, temperature, humidity, surface pressure, etc.). In this study, based on our own measurements, we present the friction conditions of an aluminum (Al 6082-T6) – polypropylene homopolymer surface pair under different surface loads (5–15 MPa) and different relative speed differences (4–8–48–96 mm/min). The results of the experiments show that the static and dynamic friction coefficients decrease with the increase of surface load. As relative speed increases, the dynamic coefficient of friction increases, while the static coefficient of friction can be considered constant.

1. Introduction

Friction is a complex phenomenon that depends on many physical and operational parameters. In addition to the material of the friction surfaces, the friction conditions between the two surfaces are greatly influenced by their surface roughness, in the case of both metals [1] [2] and ceramics [3]. The surface roughness of the material pairs has a great influence on the static coefficient of friction [4], and tribological characteristics, but the orientation of surface roughness also plays a part [5]. The effect of surface roughness generally decreases with increasing hardness [6]. Friction is also affected by humidity, and its effect is very important, e.g. in the case of brakes [7] [8], and at the same time, it also affects the entire wear mechanism resulting from friction [9]. Of course, many other parameters influence the entire friction system (e.g. ambient temperature, relative speed, surface load, etc.).

There are engineering tasks where it is necessary to know the friction conditions exactly, since it determines the operating range of the machine or working parts (e.g. form-locking joints, brakes, clutches). Nowadays, the number of material pairs used is increasing intensively, but at the same time, the relevant literature does not always provide adequate or sufficiently accurate friction data for the chosen material pairs.

As a result, several approaches and theories related to friction have been created, which Pennestrì et al. [10] described comprehensively.

If we consider friction as a mechanical system (Figure 1), the process can be divided into two stages (pre-sliding and gross sliding). In the part before sliding, adhesion forces dominate, so the friction force is primarily a function of displacement, not speed (Figure 1). The reason for this behavior may be that the contact nodes undergo elasto-plastic deformation, thus behaving as nonlinear hysteretic springs [11]. Materials included in our investigation (aluminium – AL and polypropylene – PP) materials are widely used in household and engineering applications. During use, sliding surfaces are worn out and the interfacial properties change affecting friction. This phenomenon was researched for polypropylene by Cho et al. [12]. Sedlak et al. [13] in their study was determined the changes in the sliding friction coefficient and wear mechanism of PP impregnated by oils with respect to the unmodified PP under dry sliding conditions. Their results showed that the impregnation of the PP clearly influenced the wear rate of PP and its friction coefficient. Al-Samarai et al. [14] investigated the effect of load and speed on sliding friction coefficient of aluminum–silicon casting alloy using a pin-on-disc with three different loads at three speeds. In their results they showed that the load and the speed affect the coefficient of friction.

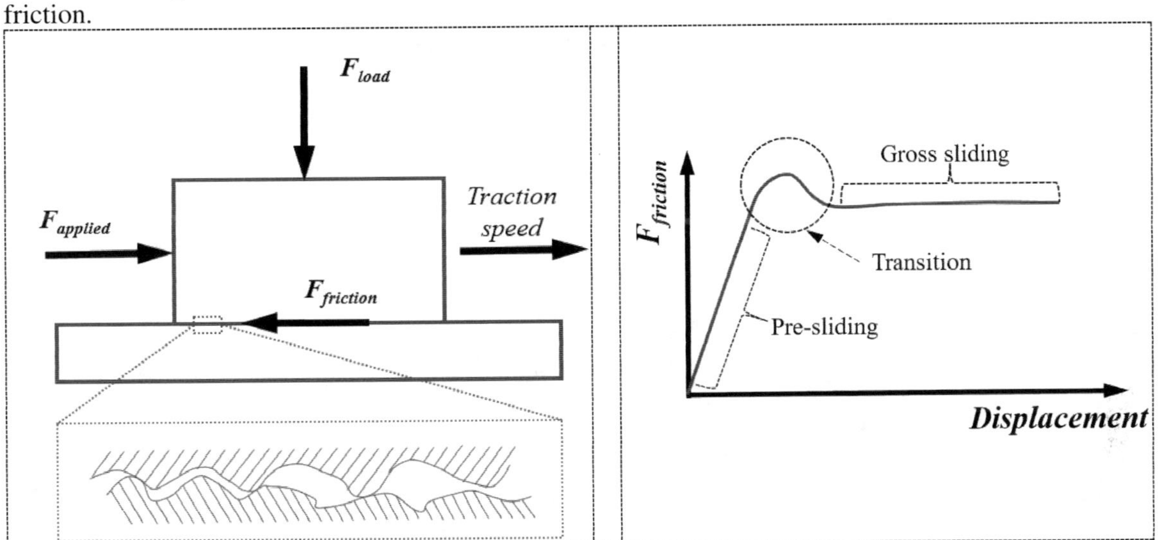

Figure 1. Basic friction configuration (left). The two friction regimes and the transition between them (right) (based on [11]).

The product $p{\cdot}v$, where p means surface pressure and v means relative movement speed, is widely used in practice to characterize the permanent load capacity and relative movement speed of pairs of materials moving on each other. Typically, the $p{\cdot}v$ value is used in two main areas of use. On the one hand, in the field of bearing materials—here this only refers to materials of sliding bearings—and on the other hand, in the field of friction materials. In addition to the product $p{\cdot}v$, it is customary to indicate the limit values p and v as well. In the case of bearing materials, the two members of the material pair constantly move on each other during operation, so the value of p is usually small and the value of v is large. In the case of friction materials—such as friction clutch and brake material pairings—the two members of the material pair do not always move relative to each other, or they only have to bear the relative movement temporarily. In this case, the value of p is typically large, and the value of v is typically small. In the present study, as part of a development task, the static and dynamic friction conditions of a pair of frictional materials had to be determined at different relative speeds and surface loads. We chose the values of p and v according to the planned range of use of the material pair. In the next chapter, we analyze the friction conditions of aluminum and polypropylene using four relative displacement speeds (4–8–48–96 mm/min) and two different surface loads (5–15 MPa). Basically, this research results show how different surface pressures affect the values of static and dynamic friction coefficient.

2. Materials and methods

2.1. Materials

The tests were performed on aluminum (Al 6082-T6) and extruded polypropylene (PP) homopolymer pairs, after the so-called run-in was performed. Run-in means initial wear, in this case 6–10 preliminary tests at the same load and the same relative speed. As a result, the transition layer resulting from the interaction of the two layers was definitely formed, and the measurement results probably characterize

IOP Conf. Series: Materials Science and Engineering **1313** (2024) 012011 doi:10.1088/1757-899X/1313/1/012011

actual operating conditions. Figure 3 show the surface structure of the samples. The aluminum samples were produced by milling (machining parameters: cutting speed, v_c = 160 m/min, feed, f_z = 0.08 mm) and the PP samples by turning (machining parameters: cutting speed, v_c = 80 m/min, feed, f = 0.05 mm). Table 1 contains the average values of the main roughness parameters of the samples before and after run-in [15].

Table 1. Main roughness parameters of the aluminum test specimen and the PP sample (before and after run-in)

	Average surface roughness *Ra*, µm	Ten-point height, *Rz*, µm	Maximum height of profile, *Rp* µm	Maximum profile valley depth, *Rv*, µm
Aluminum specimen	1.094/0.812	4.681/3.465	2.458/1.173	2.223/1.639
PP sample	0.812/0.487	3.465/2.454	1.773/0.981	1.693/1.473

Figure 2. Part of the surface of the Al 6082-T6 specimen (left), Surface of the PP sample (right)

The tests were carried out on the ~ 40–100 mm section of the aluminum specimen by dragging the Ø15 mm PP samples, using a given relative speed and a given surface pressure. During the test, we registered the $F_{friction}$ force and relative displacement with great accuracy. The test parameters were selected according to the intended use of the materials. Surface pressure was applied in the order of magnitude of the surface limit load of PP (p=5–15 MPa), this value can be determined from the material characteristics of PP. The measurement device is patented.
The static (μ_0) and dynamic (μ) coefficients of friction were calculated with the following equations:

$$\mu_0 = \frac{F_{friction\_max}}{F_{load}} \tag{1}$$

where $F_{friction\_max}$ is the maximum of the friction force (here local traction speed is close to zero).

$$\mu = \frac{F_{friction}}{F_{load}} \tag{2}$$

where $F_{friction\_max}$ is the almost steady state of the registered value (gross sliding – see Figure 2.) – traction speed is constant.

3. Results

3.1. The effect of surface load on friction

The tests were performed at a surface load of 5 MPa and 15 MPa using four speeds (4–8–48–96 mm/min). The characteristics of friction were the same in all cases. Figure 3 shows the results for the speed of 8 mm/min in detail.

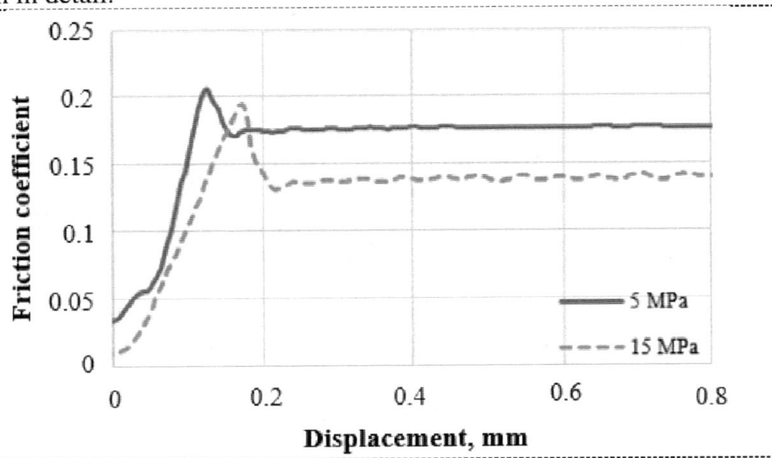

Figure 3. Registered friction ranges for different surface loads (p = 5 MPa – 15 MPa; v = 8 mm/min; frictional surface pairs: Al 6082-T6 – polypropylene homopolymer)

Figure 3 shows that the pre-sliding and gross sliding regions are almost identical even though the surface load is different, but μ_0 and μ depend on surface load. As surface load increases, both the static and dynamic coefficients of friction decrease.

It can also be observed that due to the resultant elasticity coefficient of the measuring device and the friction pairs, the rise of force (converted here to friction coefficient) is less steep at higher loads in the pre-sliding range. The measurement system was calibrated in advance.

3.2. The effect of the relative speed difference on friction

The series was also aimed at assessing the effect of the relative speed difference between the surfaces on friction. The test results are listed in Table 2. The dependence of the static and dynamic coefficients of friction on load and speed difference is shown in Figures 4 and 5.

Table 2. Static (μ_0) and dynamic (μ) coefficient of friction of the aluminum test specimen and PP sample under different surface loads and speeds

	Surface load = 5 MPa		Surface load = 15 MPa	
Traction speed, mm/min	μ_0	μ	μ_0	μ
4 mm/min	0.211	0.176	0.188	0.138
8 mm/min	0.206	0.18	0.196	0.142
48 mm/min	0.21	0.21	0.173	0.151
96 mm/min	0.23	0.24	0.18	0.163

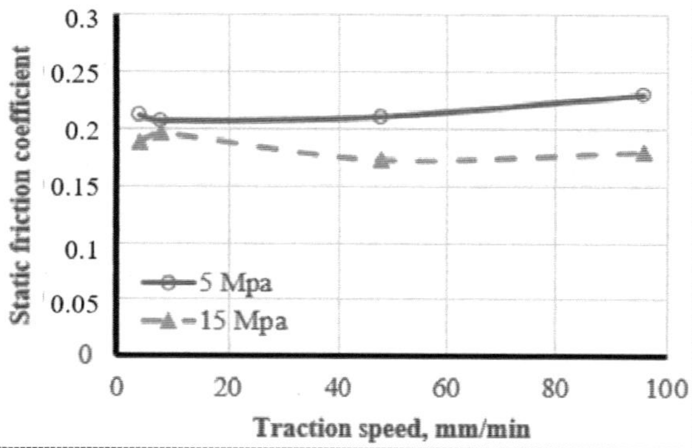

Figure 4. Static friction coefficients (surface load is 5 MPa and 15 MPa; traction speed is 4–8–48–96 mm/min; frictional material pairs: Al 6082-T6 – polypropylene homopolymer)

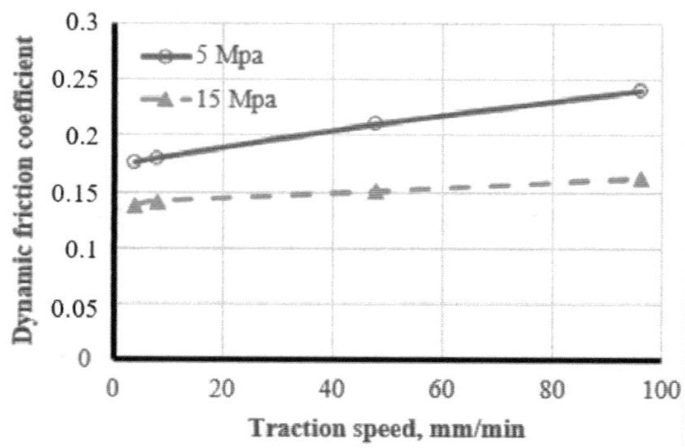

Figure 5. Dynamic friction coefficients (surface load is 5 MPa and 15 MPa; traction speed is 4–8–48–96 mm/min; frictional material pairs: Al 6082-T6 – polypropylene homopolymer)

Figures 4 and 5 show that—in the case of the studied material pair and within the applied test parameter range—increasing the surface load reduces both the static and dynamic coefficients of friction, regardless of relative displacement speed difference. Changing the relative displacement speed difference does not have a significant effect on the static coefficient of friction, it can be considered approximately constant in the test range ($\mu_0 = 0.214 \pm 0.1$, p = 5 MPa; $\mu_0 = 0.184 \pm 0.1$ p = 15 MPa). In contrast to the static coefficient of friction, the dynamic coefficient of friction depends on relative speed difference; an increase in speed difference increases the dynamic coefficient of friction in the parameter range examined.

4. Conclusion

In this study, we investigated the friction conditions of a pair of materials often used in engineering practice (aluminum, Al 6082-T6 – polypropylene homopolymer). The static and dynamic coefficients of friction depend on many parameters. We analyzed the friction conditions with two surface loads (5 MPa and 15 MPa) in the range of 4–96 mm/min relative displacement speed difference. The following conclusions can be drawn from the tests:

- both the static and dynamic coefficients of friction decrease with increasing surface load;
- the static coefficient of friction can be considered constant in the examined speed range: ($\mu_0 = 0.214 \pm 0.1$, p = 5 MPa; $\mu_0 = 0.184 \pm 0.1$ p = 15 MPa);
- increasing the surface load from 5 MPa to 15 MPa reduced the static coefficient of friction by ~ 14%;
- the dynamic coefficient of friction increases with increasing relative speed. The slope of growth decreases as surface load increases.

5. References

[1] Tsukizoe, T, Hisakado, T 1970 The influence of surface roughness on the mechanism of friction.

[2] Bachchhav B, Bagchi H 2021 Effect of surface roughness on friction and lubrication regimes. *Materials Today: Proceedings*, *38*, 169-173. https://doi.org/10.1016/j.matpr.2020.06.252

[3] Zhang W, Yamashita S, Kumazawa T, Ozeki F, Hyuga H, Kita H 2019 Influence of surface roughness parameters and surface morphology on friction performance of ceramics. *Journal of the Ceramic Society of Japan*, **127**(11), 837-842. http://doi.org/10.2109/jcersj2.19124

[4] Ghabrial S R, Zaghlool S A 1974 The effect of surface roughness on static friction. *International Journal of Machine Tool Design and Research*, **14**(4), 299-309. https://doi.org/10.1016/0020-7357(74)90019-5

[5] Jeng Y R 1990 Experimental study of the effects of surface roughness on friction. *Tribology Transactions*, **33**(3), 402-410. https://doi.org/10.1080/10402009008981970

[6] Menezes P L, Kishore Kailas S V, Lovell M R 2011 Role of surface texture, roughness, and hardness on friction during unidirectional sliding. *Tribology letters*, **41**, 1-15. https://doi.org/10.1007/s11249-010-9676-3

[7] Lee W K, Shin M W, Kim S H, Jang H, Cho M H 2013 The influence of humidity on the sliding friction of brake friction material. *Wear*, **302**(1-2), 1397-1403. http://dx.doi.org/10.1016/j.wear.2013.01.014

[8] Mirzababaei S, Filip P 2017 Impact of humidity on wear of automotive friction materials. *Wear*, **376**, 717-726. http://dx.doi.org/10.1016/j.wear.2017.02.020

[9] Chen Z, He X, Xiao C, Kim S H 2018 Effect of humidity on friction and wear—A critical review. *Lubricants*, **6**(3), 74. https://doi.org/10.3390/lubricants6030074

[10] Pennestrì E, Rossi V, Salvini P, Valentini P P 2016 Review and comparison of dry friction force models. *Nonlinear dynamics*, **83**, 1785-1801. https://doi.org/10.1007/s11071-015-2485-3

[11] Al-Bender F 2010 Fundamentals of friction modeling. In *Proceedings, ASPE Spring Topical Meeting on Control of Precision Systems, MIT, April 11-13, 2010* (Vol. **48**, pp. 117-122). ASPE-The American Society of precision Engineering; 301 Glenwood Avenue, Suite 205, Raleigh, NC 27603, PO Box 10826, Raleigh, NC 27605.

[12] Cho D H, Bhushan B, Dyess J 2016 Mechanisms of static and kinetic friction of polypropylene, polyethylene terephthalate, and high-density polyethylene pairs during sliding. *Tribology International*, **94**, 165-175. http://dx.doi.org/10.1016/j.triboint.2015.08.027

[13] Sędłak P, Białobrzeska B, Stawicki T 2016 Friction coefficient and wear resistance of a modified polypropylene impregnated with different oils. *Iranian Polymer Journal*, **25**, 263-275. https://doi.org/10.1007/s13726-016-0419-7

[14] Al-Samarai R A, Haftirman A K, Al-Douri Y. 2012 Effect of load and sliding speed on wear and friction of aluminum-silicon casting alloy. *International Journal of scientific and Research publications*, **2**(3), 1-4.

[15] Gadelmawla E S, Koura M M, Maksoud T M, Elewa . M, Soliman H H 2002 Roughness parameters. *Journal of materials processing Technology*, **123**(1), 133-145. https://doi.org/10.1016/S0924-0136(02)00060-2

Acknowledgments
Supported by the ÚNKP-23-3 New National Excellence Program of the Ministry for Culture and Innovation from the source of the national research, Development and Innovation Fund.
This research was funded by the 2020-1.1.2-PIACI-KFI-2020-00129 project. The authors acknowledge the financial support of this work by Hungarian-Japanese bilateral project (2019-2.1.11-TÉT-2020-00204.)

IOP Conf. Series: Materials Science and Engineering **1313** (2024) 012012 doi:10.1088/1757-899X/1313/1/012012

Redefining the strain aged state of unalloyed mild steels

B Kondas, V Mertinger

Institute of Physical Metallurgy Metalforming and Nanotechnolgy, University of Miskolc, H-3515 Miskolc, Hungary

E-mail: femvali@uni-miskolc.hu

Abstract. The unalloyed mild steels suitable for deep-drawing have a wide range of applications in the manufacturing industry. They are characterised by a high degree of cold formability, which is greatly reduced by the so-called "strain-aging" phenomenon. This leads to increased scrap formation, which is extremely detrimental to the metalworking industries. Strain-aging of steels is a complex subject with a very rich literature, although most of the research on unalloyed steels dates back to the 1970s and 1980s. Since then, steel making technology has developed intensively, which has also had an impact on the strain-aging process of unalloyed mild steels in practice. Our article describes the nature of the strain-aging phenomenon as it is experienced today. The degree of strain-aging described earlier in the literature is no longer fully present in mild steel products produced with today's technology.

1. Introduction

In the manufacturing industry, low carbon, unalloyed, cold-rolled, uncoated (LC) steel is the most widely used steel for cold forming. This is commonly referred to in industry narrative as mild steel. However, this steel type should not be mixed with the ultra low carbon, micro-alloyed, interstitial-free (ULC) steels, which are popular in the automotive industry and have a more specific application and a higher price than LC steels. The European standard grade for LC steels is the DC01 to 06 series, with ductility and strain-aging resistance increasing in proportion to the increase of the numbers. The basic grade DC01 is the grade where strain-aging takes place most rapidly, so that its nature can also be most clearly observed in this grade. DC01 grades are usually formed by bending, stamping, deep drawing and then welding or painting. They are mostly used to make hollow parts of various sizes, e.g. cans, dishes, drums, pipes, electrical cabinet components, household appliance covers, but are also used for producing large numbers of automotive parts. Occasionally, for economic reasons, the cheaper DC01 grade is chosen for deep drawing more complex shapes instead of the more ductile but more expensive DC03. In such cases, the increase in strength indices ($R_{p0,2}$, R_m) as a result of the strain-aging process becomes critical, along with the deterioration of elongation indices (A_{80}), especially when this occurs locally. The result: increased internal scrap, financial loss, complaints and, in more serious cases, loss of customer confidence. Strain-aging of steels, therefore, has a significant impact, especially if it is not taken into account when selecting or ordering materials. Recognising the risks, the standardisers have formulated and incorporated comments on strain-aging in the text of the standard EN 10130, which specifies the conditions for the delivery of DC01 steels. For example, in the table of mechanical properties, it is noted that the standard only guarantees the agreed yield strength (max 280 MPa) for 8 days from the date of manufacture and that Luders bands may appear on the surface of rolled sheets after a certain time. Therefore, the standard recommends that DC01 products should be processed within 6 weeks of availability **[1]**. The explanation for this is to be found in the so-called strain-aging of the steels. The strain-aging of steels has long been a well-known phenomenon and is, therefore, very well documented **[2, 3, 4, 5]**. The basic metallographic relationships were established as early as in the 1930s and 1950s. Cold rolled sheet for deep drawing was a popular product of the post-World War II re-industrialisation, so a great deal of scientific research on strain-aging was carried out in the 1960s and 1970s. The currently available literature is predominantly concerned with the results of products made with now obsolete technology. In the last decades, steelmaking technologies have undergone a major development which has resulted in an extraordinary improvement in the quality of rolled steel products. In this article, we will investigate the nature of strain-aging of DC01 sheets produced with current state-of-the-art technology and compare the results with those reported in the literature, and draw conclusions.

1.1 The causes and manifestations of strain-aging

1.1.1 Static strain-aging

Formational strain-aging is the result of a process of dislocation by movement and diffusion. During this process, the alloying atoms (carbon and nitrogen) and impurity elements in solution in the material migrate to the end of the dislocation, where they cause less lattice strain. Diffusion of these small and mobile atoms can take place in a practically perceptible time at room temperature. As time passes, the amount of atoms reaching the dislocations increases (forming a Cottrell cloud), which makes the dislocations increasingly difficult to move. When the forming of sheets is started, the dislocations will break away from these atoms, but this requires a higher stress value. Once the dislocations have broken away, a smaller tension is sufficient to move them further, which means a reduction in the measured force. This causes the appearance of the upper and lower yield strengths **[6, 7,8]**.

1.1.2 Dynamic strain-aging

The Portevin-Le Chatelier (PLC) effect (jerky flow) does not occur at the elastic-plastic behaviour interface, as in static strain-aging, but well afterwards in the strain hardening region. The PLC effect is caused by the dynamic interaction of moving dislocations and rapidly diffusing alloy atoms. In this case, the dislocation motion is not continuous but mostly intermittent. After the dislocations break away, they are slowed down by new obstacles until they eventually stop again. During the waiting time, the solute atoms around the dislocations segregate, which represents a steadily increasing strength of the obstacle with waiting time. Thus, on the one hand, the strain-aging effect of the moving alloying atoms, and on the other hand, the separation of the moving dislocations from the collection of solute atoms, facilitated by thermal activation, give rise to the competing processes that alternate and constantly recur. The process is characterised by the fluctuation of stress in the stress-strain curve. In terms of PLC band propagation, Kim et al. have defined different types of motifs **[9,10]**. As a result of the above metallographic processes, the yield strength and tensile strength are increased, but the elongation is reduced and Lüders bands appear on the surface. This finding has been described by many researchers as a result of their experiments. Sylwestrowicz and Hall **[11]** found a yield strength increase of about 70 MPa in mild steel sheets after a storage time of about 6 weeks at room temperature, Hrvnak and Sobotova **[12]** measured a 22 MPa increase and a 6% decrease on sheets of the same age, while Bhagat **[13]** measured an 18 MPa increase and a 0.6% yield strength decrease.

1.2 Factors affecting the measurement of strain-aging.

1.2.1 The test speed

The large difference between the results of the three researchers may be due to the different forming speeds, i.e. different tensile test fracture rates, which significantly affect the results of mechanical properties. Jocham et al. observed a difference of about 50 MPa between the results of 0.11/s and 0.00025 1/s strain rate **[14]**.

1.2.2 The test temperature

Temperature is the most obvious cause of strain-aging due to diffusion. The effect on both static and dynamic strain-aging has been investigated by Brindley and Barnby. Strain-aging also occurs under normal industrial conditions (-10 to +30 ºC), although much more slowly than at elevated temperatures **[15]**.

1.2.3 Other factors

In order to accurately measure the change in mechanical properties during strain-aging, it was considered necessary to assess the effects on the tensile test results. The analysis was carried out using the Ishikawa method and the results are shown in **Figure 1.**

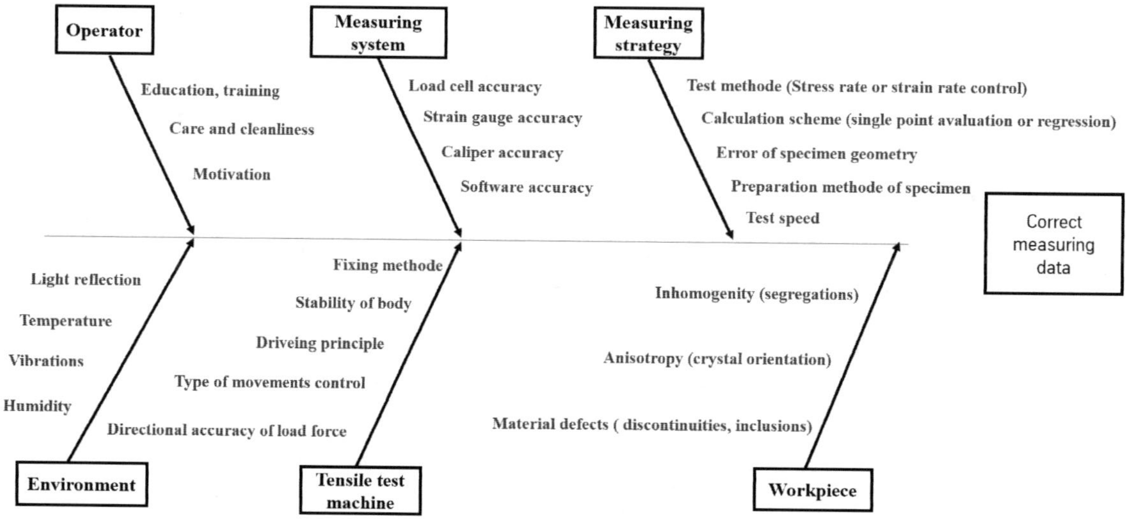

Figure 1. Factors affecting tensile test results

2. Experimental

Based on the literature, it has been found that with strain-aging the following changes in the properties of mild steels occur:

- Increasing yield strength and tensile strength
- Reduced elongation after fracture
- The upper and lower yield strength (static strain-aging) appears
- A serration appears in the uniform elongation section of the stress-strain diagram (dynamic strain-aging)

My research has, therefore, focused on identifying these four strain-aging effects.

2.1 Material

The effects of strain-aging were analysed on standard cold rolled, annealed and temper-rolled samples of DC01 grade from a European manufacturer, using the results of tensile tests made 3 days, 9 days, 2 weeks, 1 month and 6 weeks after production **(Figure 2)**.

C%	Mn%	Si%	S%	Al%	N%
0,07	0,39	0,007	0,001	0,038	0,004

Figure 2. Chemical composition of investigated sheets

The time intervals were determined according to the relevant product standards. The tensile test specimens were cut from ¼ of the width of the sheets from the third turn of the coil, closely spaced to each other, in order to keep the inhomogeneity of the material as low as possible **(Figure 3)**. Three specimens were tested per series (0, 45 and 90 degrees to the rolling direction). The tensile tests were performed on specimens of standard size (ISO 6892-1, Type B (20X80mm)), prepared for testing by punching and grinding.

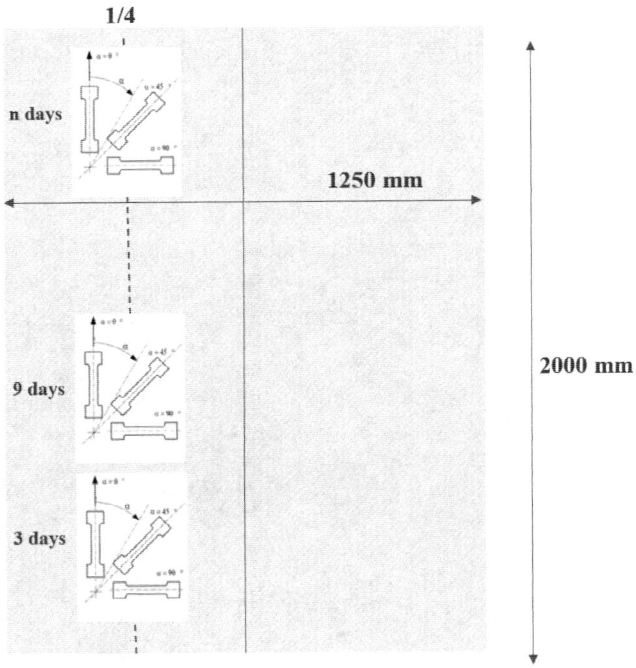

Figure 3. Taking specimens in the sheets

2.2 Testing machine and method

The tensile tests were performed on a dual column floor model universal material tester type Instron 68FM-300. Measurement results were processed using the evaluation software type Bluehill 3. During the test, the elongation of the specimens was measured using an Aver Edge 32 video extensometer with a signal spacing of 80 mm longitudinally and 20 mm transversely. Additional parameters of the test; Load cell: 300 kN; Test temperature: room temperature (23°C). The variation of the mechanical values ($R_{p0,2}$, R_{eH} , R_{eL} , R_m , A_{80}) was tested per series using the ISO 6892-1 Method B (stress rate control) on three test specimens per direction. The crosshead separation rate was uniformly 10 MPa/s up to the appearance of the yield point and 0.00671 1/s thereafter. Static and dynamic strain-aging was tested using the ISO 6892-1 Method A1 (strain-rate control) with a tension rate of 1-1 test per direction in three tests (0.002 1/s; 0.0002 1/s; 0.00002 1/s). All influencing factors described in section 1.2.3 were kept constant during the experiments. The same data acquisition frequency was used for all tensile tests. The bias of the measurement system was checked by testing certified standard specimens before each series of measurements.

3. Results and Discussion

3.1 Results of the testing of mechanical values

The first tensile test was carried out on the 3rd day after production, where we were able to confirm the values on the manufacturer's test certificate of material supplied with the consignment. The variations of the mechanical properties over time are shown in **Figures 4, 5 and 6**. The results show that no clear trend could be detected for either the yield strength, tensile strength or elongation after fracture over the time interval tested. The changes were small, corresponding to a combination of measurement uncertainty and material inhomogeneity.

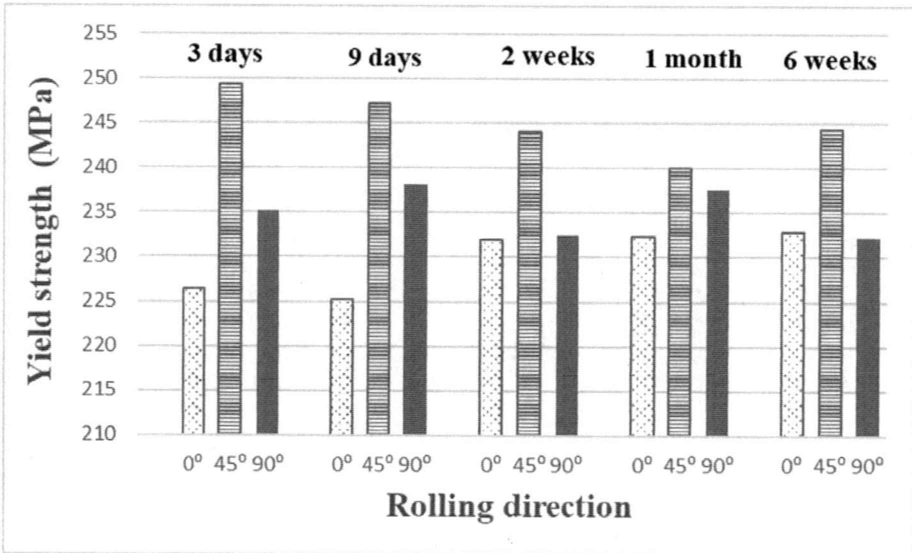

Figure 4. Variation of yield strength with time in different directions

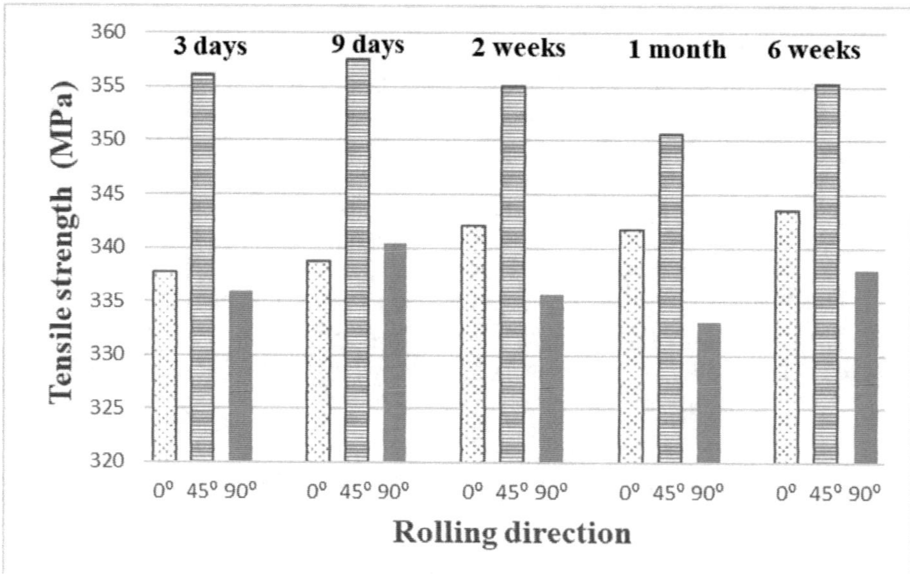

Figure 5. Tensile stress changes with time in different directions

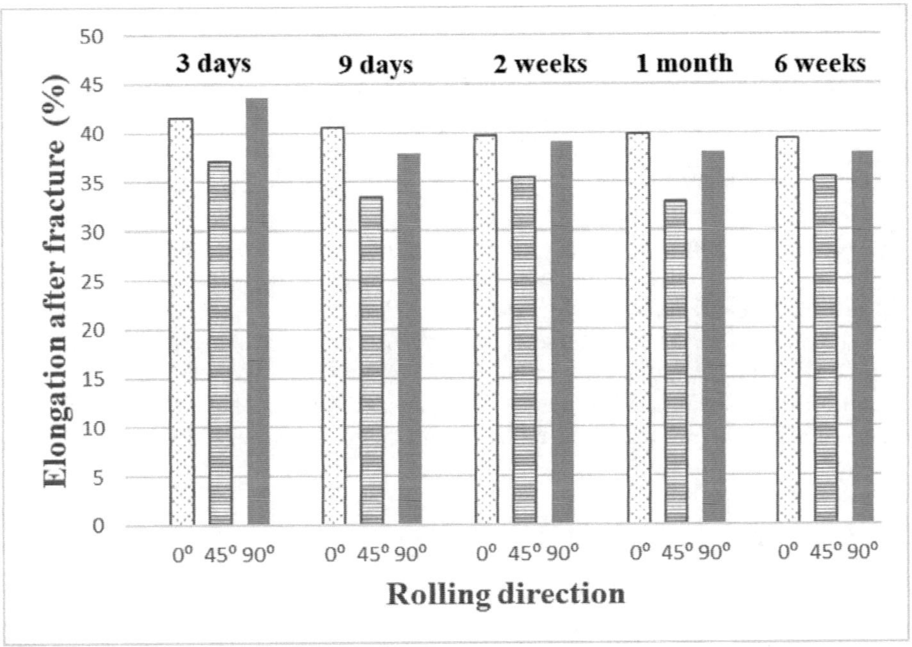

Figure 6. Change in elongation after fracture with time in different directions

3.2 Results of the testing of static and dynamic strain-aging

One of the most surprising results of the series of tests was that no lower or upper yield strengths could be determined for any of the 108 tests, hence, no static strain-aging could be detected. The effect of dynamic strain-aging was tested during the uniform elongation phase of the tensile curves, always at the same elongation. The results are shown **in Figure 7**. The results show that the so-called serration, which is characteristic of dynamic strain-aging, was only clearly detectable at a strain rate of 0.00002 1/s and that it only reached dimensions that could be evaluated without magnification at 6 weeks of age. At the higher strain rate, however, there was no detectable serration at all, even on the diagrams taken using the method testing the mechanical properties (see ISO 6892-1 Method B)!

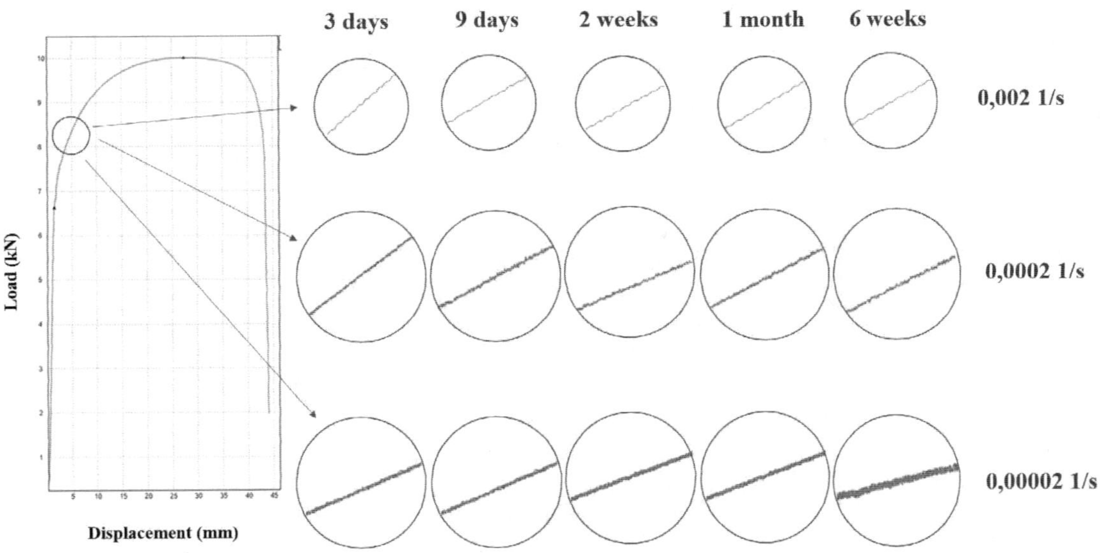

Figure 7. Signs of dynamic strain-aging over time at different strain rates

IOP Conf. Series: Materials Science and Engineering **1313** (2024) 012012
doi:10.1088/1757-899X/1313/1/012012

3.3 Results of Erichsen and Deep-drawing cupping tests

As noted earlier, the metallographic process of strain-aging results in a reduction in the ductility of steel materials. In order to verify that strain-aging has indeed taken place in our samples, or at least that the essential part of it has taken place, we considered it necessary to carry out technological tests. At the time of each series of tensile tests, we also took specimens for the Erichsen and Deep-drawing cupping tests next to the tensile test specimens from the same sheet. For the deep-drawing cup tensile test, blanks of progressively increasing diameter were cut and drawn into cups by means of a pulling die of a given rounding and the same diameter. The diameter of the blank that can still be cupped without fracture is considered as the property of deep-drawing (IG). The Erichsen cupping tests were performed and evaluated according to EN ISO 20482. The results of the tests are shown in **Figures 8. and 9**. The results show that both the IE and IG values decrease with time, indicating a degradation of ductility. There is a steady decrease in Erichsen's value (IE) between the 3-day and 6-week samples, while there is a negative change in the maximum diameter that can be drawn into a cup (IG) after 9 days and 6 weeks.

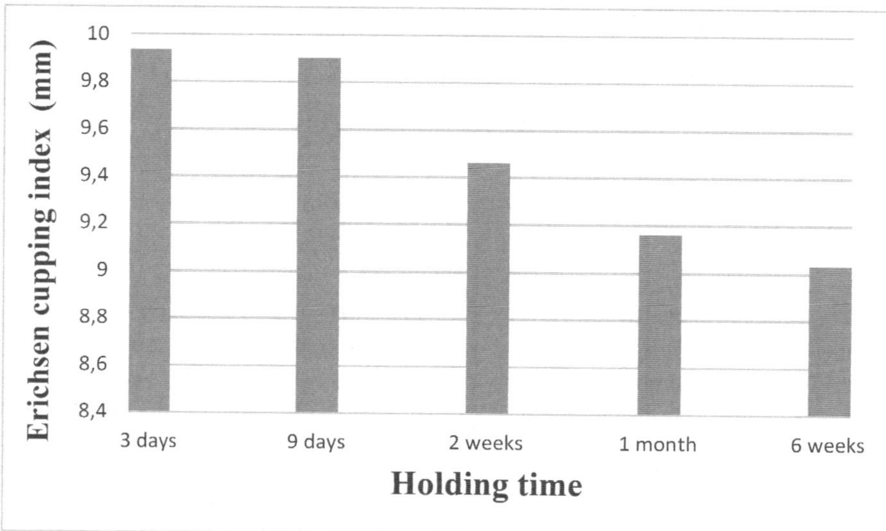

Figure 8. Erichsen values as a function of elapsed time

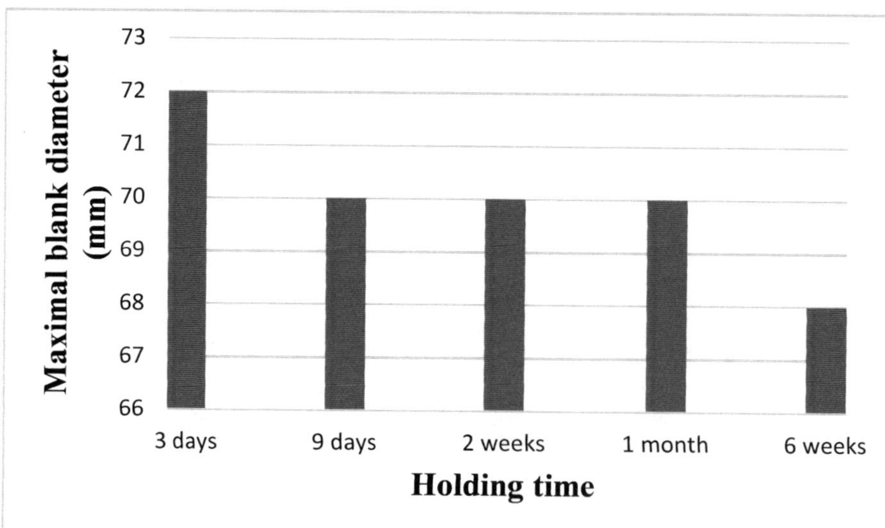

Figure 9. Cupping test results as a function of elapsed time

IOP Conf. Series: Materials Science and Engineering **1313** (2024) 012012 doi:10.1088/1757-899X/1313/1/012012

4. Conclusion

Degradation during the strain-aging of steels is a process for which there is no specific indicator, but several types of strain-aging symptoms are distinguished in the literature. In this paper, the effects of strain-aging were tested as described in the literature on mild steel specimens produced with today's modern steelmaking technology. Based on the tensile tests of our experiments, we could not detect static strain-aging, i.e. the formation of upper and lower yield strengths, in the time interval investigated, and dynamic strain-aging, i.e. the serration, only after 6 weeks and then only at an extremely low strain rate (0.00002 1/s). Increases in yield strength and tensile strength values and decreases in elongation after fracture values were not observed in the tested specimens either. The fact that the ductility was nevertheless affected by the time progression was verified by conducting technological tests during our experiments. The critical age was found to be 6 weeks, where the onset of dynamic strain-aging was accompanied by a drastic reduction in the blank diameter. The results suggest that it would be particularly useful for industrial applications to define precisely the stage of the strain-aging process at which the possibility for forming deteriorates to a critical degree. As it was seen, the degree of strain-aging described earlier in the literature is no longer present in mild steel products produced with today's technology. This is due to the computer controlled desulphurisation, vacuum refining and continuous casting technologies used today **[16, 17, 18, 19]**. The development of a complex indicator(s) for the determination of strain-aging that can also be used conveniently by industry will be the next stage in our research.

5. References

[1] ECISS/TC 13 2007 Cold Rolled Low Carbon Steel Flat Products for Cold Forming Technical Delivery Conditions *DIN EN 10130*

[2] Bhagat A N, Baek S, Lee H C 2008 A Simple Method for Prediction of Shelf Life of Bake Hardening Steels ISIJ International **48** 1781-1787

[3] Gosh P, Ray R K 2017 Deep Drawable Steels Automotive Steels, Design, Metallurgy, Processing and Applications, 113-143

[4] Sarkar B, Datta R, Chaudhuri S K 2009 Effect of Titan Nitrogen Ratio on Strain Aging and Plastic Anisotropy in Extra Deep Drawing Quality Steel Materials Science and Technology **25** 1408-1410

[5] De A K, De Cooman B C, Vandeputte S 2001 Kinetics of Strain Aging in Bake Hardening Ultra Low Carbon Steel Journal of Materials Engineering and Performance **10** 567-575

[6] Cunningham S 1999 Effect of substitutional elements on dynamic strain aging in steel *McGill University* 101

[7] Cottrell A H and Bilby B A 1949 Dislocation Theory of Yielding and Strain Ageing of Iron *Proceedings of the Physical Society Section A* **62** (1), 49–62.

[8] Cottrell A H 1948 Report on the Strength of Solids *London Physical Society* **30**

[9] Penning P 1972 Mathematics of the Portevin-Le Chatelier Effect *Acta Metallurgica*, **20** 1169–1175

[10] Campbell J D 1953 The Dynamic Yielding of Mild Steel *Acta Metallurgica* **1**

[11] Sylwestrowicz W and Hall E O 1951 The Deformation and Aging of Mild Steels *Proc. Phys. Soc. B* **64** 742

[12] Hrivnak A and Sobotova L 1992 The Infulance of the Deformational Aging and the Condition of Stress on the Properties of the Deep-Drawing Steel Sheet *Journal of Materials Procesing Technology* **34** 425-430

[13] Bhagat A N, Baek S J and Lee H C 2008 A Simple Method for Prediction of Shelf Life of Bake Hardening Steels *Material Science Engineerings* **48** 1781-1787

[14] Jocham D, Norz R and Volk W 2016 Strain Rate Sensitivity of DC06 for High Strains under Biaxial Stress in Hydraulic Bulge Test and Under Uniaxial Stress in tensile Test International *Journal of Material Forming* **10** 453-461

[15] Brindley B J and Barnby J 1966 Dynamic Strain Ageing in Mild Steel *Materials Science Acta Metallurgica* **12** 1765-1780

[16] Fandrich R, Lüngen H B, Wuppermann C D 2007 Secondary metallurgy, status 2007 *7th Clean Steel Conference* 28-37

[17] Károly Gy, El-Ghazaly, Harcsik B, Szabo A 2009 The Effect of Flow Induced by Argon Stirring on the Cleanliness of Low Silicon, Aluminium Killed Steels *Conference on Modelling Fluid Flow CMFF'09* **937** 580-585

[18] Karoly Gy, Tardy P 2007 Steel Technologies in the Changing World of Steel, Historical Review and Recent Developments *Metals Processing and Manufacturing Conference* CD 6p

[19] Ito K, Shioya M, Ago M, Kurokawa T, Kobayashi H, Mori J 2019 Production Planing and Scheduling Technology for Steel Manufacturing Process *Nippon Steel & Sumitomo Metal Technical Report* **121**

IOP Conf. Series: Materials Science and Engineering **1313** (2024) 012013 doi:10.1088/1757-899X/1313/1/012013

The challenges of 3D metal printing and the machinability characteristics of additively produced materials: a review of hybrid machining

B Z Balazs[1,*], A Jacso[1], V Tancsa[1], and M Takacs[1]

[1] Department of Manufacturing Science and Engineering, Faculty of Mechanical Engineering, Budapest University of Technology and Economics, H-1111 Budapest Műegyetem rkp. 3., Hungary

[*] Corresponding author: balazs.barnabas.zoltan@gpk.bme.hu

Abstract. Recently, 3D metal printing technologies are rapidly spreading. The various types of additive manufacturing (powder bed fusion, directed energy deposition, binder jetting, and sheet lamination) use very different approaches to produce materials/geometries. The material properties (e.g. porosity/density, strength, surface roughness, hardness) of additively manufactured materials differ from the traditional ones and strongly depend on the applied 3D printing parameters. Therefore, a crucial research direction is how the parameters affect the material properties. Currently, 3D metal printing technologies have limitations in the achievable geometrical accuracy and surface roughness. Therefore, they can rarely be used for part production without further machining. For high-quality requirements, an additional subtractive machining procedure is usually required. Consequently, another intensively researched topic is the machinability of 3D metal printed materials. The industry is increasingly demanding to integrate different manufacturing processes (additive and subtractive technologies) into one machine. These hybrid machines have the advantage that they can produce high-quality parts in one step.

1. Introduction

Over the past decades, additive manufacturing (AM) technologies have developed rapidly. As a result, 3D printing of a wide range of materials, such as polymers, metals, ceramics, composites, biomaterials, concrete, etc., has become available [1]. According to Wohlers Report 2023, the industry revenue related to additive manufacturing tripled from approximately $6 billion in 2016 to $18 billion in 2022 and is expected to exceed the $100 billion mark by 2032 [2]. Although polymer materials are still dominant in 3D printing, the additive manufacturing of metals is also catching up at an accelerating pace [3]. Precedence Research reports that metal 3D printing had a $2.9 billion market in 2022 and is expected to grow fivefold by 2032 [4]. In the following, this paper aims to focus only on this area.

The development of the first metal 3D printing process is attributed to J. F. Gottwald (AT&T Teletype Corp), who patented a jetting-based free-form object printing process in 1969 [5], [6]. However, technologies that can satisfy industrial needs only appeared around the turn of the millennium [7]. Additive manufacturing of metals offers excellent freedom in construction design, which provides an advantage in many areas over traditional forming and subtractive manufacturing technologies [8]. However, even with the leading additive technologies, achieving the same dimensional accuracy and surface quality indicators as with traditional technologies is impossible. Therefore, a finishing process is usually required on the functional surfaces [9]. If strict dimensional accuracy requirements must be met, the finishing process can be performed by cutting, usually by 3-5-axis milling [10]. In most cases, this is done on a separate machine tool, but there is an increasing demand for hybrid machine tools where the part can be manufactured in one clamping [11]. This approach can provide the benefits of both additive and subtractive techniques, enabling the production of complex geometries with improved efficiency and precision.

Hybrid additive–subtractive manufacturing can be divided into two process patterns depending on timing: sequential and iterative [12]. In the case of sequential order, the additive manufacturing phase is followed by the subtractive phase as a final operational element. In contrast, in the case of iterative order, the two technologies follow each other alternately. Although the implementation of sequential order is significantly simpler, the iterative approach has the advantage that the surfaces to be machined are more easily accessible during the construction of the 3D geometry [13]. Nowadays, innovative solutions appear in large numbers in both fields.

Although additive manufacturing can be beneficial in many areas, an entirely new attitude to design and technology is required to fully exploit its potential. Accordingly, adopting the Design for Additive Manufacturing approach is considered a vital issue [14]. To review these aspects, the following part of the paper briefly summarizes the methods of additive manufacturing of metals (Section 2-4) and the machinability aspects of these materials (Section 5).

2. 3D metal printing

3D metal printing technology can be classified into four main groups: Powder Bed Fusion, Directed Energy Deposition, Binder Jetting, and Sheet Lamination [15].

The Powder Bed Fusion (PBF) technology is an extensively researched and widely applied additive manufacturing technology. Its essence lies in melting the powdered material layer-by-layer using thermal energy. The process is basically a nonequilibrium solidification process. The heat source depends on the specific technology and can be either an electron beam (EBM) or a laser (LPBF – laser powder bed fusion, SLS – selective laser sintering, SLM – selective laser melting, DMLS – direct metal laser sintering) [16,17]. Its primary applications are in the aerospace and automotive industries, and the manufacturing of medical implants and customized parts [18]. This technology provides high precision and strength in complex geometries, and excellent material properties [19]. The main limitations are the smaller build size compared to other technologies, and usually, post-processing is required for surface finish [20].

Directed Energy Deposition (DED) is another significant subset of additive manufacturing that involves depositing material, either in powder or wire form, onto a specific area. The process gradually builds the desired geometry layer-by-layer through localized melting of the material using a heat source such as a laser, electron beam, or plasma arc. DED methods vary based on the type of material used (powder or wire) and the energy source applied (laser, electron beam, and arc) [16,21]. Therefore, the technologies can be distinguished into three groups according to the energy source: direct laser deposition (DLD), electron beam additive manufacturing (EBAM), and wire arc additive manufacturing (WAAM). In LMD, laser-melted metal powder is used for building, while LMD-W uses a wire-based laser metal deposition. The WAAM process uses an electric or plasma arc for heat, introducing wire as the construction material [22]. The primary applications of DED technologies are repairing and re-manufacturing components and large-scale metal part production. This technology provides high deposition rates for rapid builds and is excellent for repairing or adding material to existing components [23]. The main limitation of its application is the lower precision compared to some PBF methods, and usually, post-processing is needed to improve dimensional accuracy [24].

At Binder Jetting (BJ), a liquid binder is precisely deposited onto the metal powder layer using a print head. The binder binds the metal powder particles together in the desired pattern corresponding to the cross-section of the 3D object, and the process is carried out layer-by-layer [25]. The primary applications of BJ technology are rapid prototyping, low-volume production and fabricating sand moulds and cores for the metal casting industry [26]. This technology provides high printing speed and enables the production of large and complex metal parts without support structures [27], but with lower resolution and density, usually, heat treatment and other post-processing steps may be required for strength and durability [28].

In the case of Sheet Lamination (SL), also known as Laminated object manufacture (LOM), the desired object is created by bonding together thin sheets [29]. In the case of metals, ultrasonic vibrations are commonly utilized (Ultrasonic Additive Manufacturing, UAM), meaning the bonding is achieved through ultrasonic welding [30]. The primary applications of SL are prototyping, concept modelling, producing composite materials and large-scale metal part production [31]. This technology provides cost-effective and rapid manufacturing of large parts and versatility with different sheet materials [32]. Additionally, it is the only method capable of creating metal structures at relatively low temperatures, allowing embedding sensitive components like printed circuits between layers [33]. The main limitation of its application is the low resolution characterized by poor surface quality and dimensional accuracy, and the removal of excess material after bonding is often challenging, especially for complex shapes [34].

Table 1. Overview of 3D metal printing techniques.

Technology	Fundamental principle	Main categories	Advantage	Disadvantage
PBF	A thermal energy fuses a small designated region of the powder bed to build components	- EBM - LPBF, SLS, SLM, DMLS	- High accuracy - High mechanical strength	- Relatively slow process - High-cost equipment - High-cost materials
DED	A focused thermal energy melts metallic materials during deposition	- DLD - EBAM - WAAM	- Large components - Low manufacturing time and cost - Using multiple materials simultaneously	- Low accuracy and surface quality - Low feature details
BJ	Metal powder binding, followed by sintering process	-	- High speed - High complexity without support	- Low resolution - Low density
SL	Sheets of materials selectively stack and consolidate into products	- UAM	- High speed - Low cost - Using multiple materials simultaneously	- Weak mechanical properties in the build direction - Requirement of postprocessing

After a brief review of the most common 3D metal printing technologies (table 1.), the rest of the paper focuses on PBF technologies, paying special attention to SLM and DMLS. These technologies were chosen because they have the most significant industrial potential due to the best accuracy and mechanical properties.

3. Difficulties with 3D metal printed materials

During additive manufacturing, the base metal powder undergoes a significant structural transformation compared to the conventional block material. These differences are manifested in many material properties [35].

Rapid heating during the PBF process can result in significant residual stresses within the material. These stresses may affect the part's dimensional accuracy, warping, and mechanical properties. To manage residual stress, post-heat treatment is usually unavoidable [36]. The materials produced by PBF processes can often be characterized with highly anisotropic mechanical properties due to the layer-by-layer construction. Strength, ductility, and fatigue resistance may vary in different directions, influencing the component's overall performance [37]. The layer-by-layer construction of material can also lead to microstructural irregularities, such as variations in grain size and grain boundary structure. These irregularities may impact material strength, hardness, and other mechanical properties.

Incomplete fusion of powder particles or gas entrapment during the melting process can produce porosity within the printed material [38]. Porosity can compromise mechanical properties, reducing tensile strength and fatigue resistance [39]. Contamination of the metal powder or the construction environment can add undesirable particles to the material, affecting its purity and potentially leading to variations in mechanical properties [40]. Variations in layer adhesion, laser power, or powder distribution can also lead to inconsistencies in the final products [41]. Furthermore, the thermal cycling involved in the PBF process can create a heat-affected zone with altered microstructure and mechanical properties. Managing heat-affected zone effects is essential to ensure consistent material properties [42].

4. Effect of the printing parameters on the material properties

The actual material properties strongly depend on the applied 3D printing parameters. This section concentrates on the influence of the printing parameters on the material characteristics in the case of laser PBF technologies. In the case of SLM, there are many adjustable technological parameters, e.g. laser power, scanning speed, hatch spacing, hatch scanning angle, layer thickness, laser diameter, building direction, and energy density (it is dependent on the other parameters). The key material properties are e.g. material density/porosity, surface quality, tensile strength, hardness, and grain size.

Tucho et al. [43] investigated the effect of the 3D metal printing parameters, such as scanning speed, hatch spacing, energy density on the material density, material defects (e.g. porosity), and its micro-hardness on SS316L steel. In their study, the energy density was the most influencing parameter on the properties. The average porosity was generally low, less than 4 % in the range of 40-60 J/mm^3. However, the porosity exponentially decreases with the increasing energy density. Applying 80 J/mm^3 energy density resulted in 99.8% material density. It was also found that the higher hatch spacing caused more pores and lower material density. Therefore, using small hatch spacing combined with higher scanning speed is recommended to reach high material density. Lu et al. [44] investigated the influence of SLM parameters on NiTi shape memory alloys. Their experiments showed that the highest density was not necessarily achieved by the highest laser energy density. An optimal parameter combination (70 W laser power and 105 mm/s scanning speed) was found for the highest relative density. Therefore, it is recommended to consider the parameters' combined effect. Zhang et al. [45] also investigated the density of the workpiece material at the different hatch scanning angles (0°, 15°, 30°, 45°, 60°, 75°, 90°, where 0 is parallel to the longitudinal axis of the tensile test specimen). The density values were in the range of 95.2 to 95.9% at the SLM-manufactured 316L material. The highest value was found at a 30° hatch scanning angle, and the lowest was at 45°. The higher density means that there are fewer defects in the material. Nath et al. [46] investigated the influence of layer thickness using 420 stainless steel material and laser powder bed fusion technology. Using 10 μm or 20 μm layer thickness, it was found that the material density is higher than applying 30 μm.

Tensile strength is another important mechanical property of the workpiece material. Based on scientific and practical investigations, it is also strongly influenced by the printing parameters. Zhang et al. [45] investigated the effect of the placement orientation at SLM-manufactured 316L, and it was found to influence the workpiece's tensile strength greatly. The highest value was found when the fabrication direction was orthogonal to the load. This can be because the adhesive properties between two adjacent melting paths are stronger than between two adjacent layers. Moreover, the different hatch scanning angles were also investigated in the range of 15-90°. It was found that the optimal angle at each layer is 30°. Nath et al. [46] proved experimentally that the layer thickness significantly influences the material's tensile strength. The smaller the layer thickness, the higher the tensile strength. At 10 μm layer thickness, the ultimate tensile strength was 1130 ± 35 MPa, while in the case of 30 μm layer thickness, the tensile strength was only 760 ± 35 MPa.

The applied 3D metal printing parameters also influence the material hardness; moreover, the hardness values strongly correlate with the occurred porosity. The smaller the porosity (or higher the material density), the higher the material hardness [43,45]. Tucho et al. [43] discovered that energy density can be an appropriate parameter to control the hardness of the material since the hardness increases linearly with the increasing energy density at SS316L. The hardness was 176 ± 9 HV at 50 J/mm^3 and 213 ± 3 HV at 80 J/mm^3. However, the workpiece has very significant anisotropy. The difference was 14% between measuring the hardness in the surface-normal and the surface-parallel of surface. Ni et al. [47] experienced the same phenomena on SLM-manufactured Ti6Al4V. Al-Rubaie et al. [48] found that the SLM-produced material (Ti6Al4V) is 20% harder than the conventionally made material. Zhang et al. [45] investigated the hatch scanning angles (0°, 15°, 30°, 45°, 60°, 75°, 90°) and it was found that the highest micro-hardness value belongs to the 30° hatch scanning angle, and the lowest value was measured at the 45° angle. Nath et al. [46] found that the hardness of the material is influenced by the layer thickness, too. The value was 57 ± 1 HRC at 10 µm layer thickness, and this value was decreased to 51 ± 1 HRC in the case of using 30 µm layer thickness at 420 stainless steel. As mentioned in Section 2, the SLM process is a nonequilibrium solidification process, and depending on the material cooling rate, different phases can occur, affecting the hardness. In the case of traditional materials, there is a wide range of information about the effect of grain sizes, which knowledge could be adapted to additively manufactured materials. The grain size of the workpiece material also has a significant influence on the properties as well as the machinability of the material. The movement of dislocations is inhibited by the grain boundaries. Based on this, the dislocation movement requires more energy because the larger grain boundary length has a more inhibiting effect in the case of smaller grain sizes. The cutting processes (e.g. the micro-milling) are partly based on the dislocation moving. Therefore, a material structure with smaller grains necessarily means higher cutting forces during machining [49]. This fine-grained structure limits dislocation movement, making the base material more resistant to deformation. The reduced deformability occurs with the increase in hardness [50,51]. Using smaller-grain-sized material, the machining is performed by greater cutting forces, but the machining accuracy can be improved [52].

5. Machining of SLM-printed metals

Currently, the additively manufactured parts cannot meet the high-quality requirements of surface roughness and geometric accuracy. Therefore, manufacturing a part usually involves a subsequent subtractive machining process (e.g., milling and micro-milling). Consequently, it is also necessary to investigate the cutting process in detail due to the very different properties of the additively produced materials [48,51]. Le Coz et al. [51] investigated the cutting forces during micro-machining of Ti6Al4V. It was found that cutting forces were higher (3-24%, depending on the cutting conditions) in the case of SLM-produced material than the traditional cast material. Al-Rubaie et al. [48] experienced the same phenomena in the case of conventional-sized milling of Ti6Al4V material. Polishetty et al. [53] also found similar characteristic at thread milling of wrought and SLM-made 316L material. Shunmugavel et al. [54] compared the machinability of the SLM-manufactured Ti6Al4V with the traditionally produced one. It was observed that SLM material has higher yield and ultimate strength and hardness, but its ductility is worse. Therefore, the cutting forces were also higher at SLM-produced material. The cutting speed at the traditionally made material can reduce the cutting forces due to thermal softening. However, in the case of SLM-manufactured workpieces, the presence of porosity inhibits it. Furthermore, using SLM material, better surface roughness can be produced due to its hardness and more brittle behaviour. Ni et al. [47] investigated the effect of laser scanning strategies (0°, 67.5°, and 90°) on the machinability of Ti6Al4V. The anisotropy of the workpiece material results in differences in the cutting force when machining different sides of the part. The surface roughness was significantly better on the top surface in the case of 0° and 90°, but applying 67.5°, the front surface roughness is slightly better. The connection between roughness and feed rate differs in the top and front surfaces. Based on the study by Al-Rubaie et al. [48], despite the different material properties (e.g. residual stresses and higher hardness) the machinability of SLM material does not show significant differences. According to the authors, the conventionally applied cutting parameters can be adapted to the SLM-made materials.

6. Summary and outlook

The 3D metal printing technologies are spreading more and more widely. Currently, there are many procedures for additive manufacturing of materials. With the advance of 3D metal printing, more and more manufacturers offer machines and develop these technologies. One of the most important additive manufacturing technologies is the SLM/DMLS since it can produce relatively dense material with relatively good precision and quasi-good surface roughness, and the material homogeneity is also appropriate. The material properties are strongly dependent on the applied 3D metal printing parameters. Usually, additive manufacturing technologies are not suitable for producing finished parts. Therefore, subsequent subtractive machining is required. Thus, the other current and intensively researched topic is the investigation of the machinability of 3D-printed metals. The machining characteristics show many differences between the additively and the traditionally made base material. However, based on the scientific literature, the cutting parameters can be well-adapted to the 3D-printed metals despite the differences in material properties.

Hybrid manufacturing technologies in metal part production represent a promising future. Combining the advantages of additive and subtractive techniques in a single system offers enhanced flexibility, precision, and efficiency, enabling the production of complex and high-precision metal components. However, a real technological breakthrough can only be expected if considering the specifics of additive manufacturing also extends to structural design since fully exploiting the potential inherent in this technology is only possible in conjunction with well-thought-out structural and technological planning.

7. References

[1] Azlin M N M, Ilyas R A, Zuhri M Y M, Sapuan S M, Harussani M M, Sharma S, Nordin A H, Nurazzi N M and Afiqah A N 2022 3D Printing and Shaping Polymers, Composites, and Nanocomposites: A Review *Polymers* **14** 180

[2] Wohlers Associates 2023 *Wohlers report 2023: 3D printing and additive manufacturing global state of the industry* (Washington (D.C.): Wohlers Associates)

[3] Madhavadas V, Srivastava D, Chadha U, Aravind Raj S, Sultan M T H, Shahar F S and Shah A U M 2022 A review on metal additive manufacturing for intricately shaped aerospace components *CIRP J. Manuf. Sci. Technol.* **39** 18–36

[4] Anon Metal Additive Manufacturing Market Size, Share, Report 2023-2032

[5] Ansell T Y 2021 Current Status of Liquid Metal Printing *J. Manuf. Mater. Process.* **5** 31

[6] Gottwald J F 1971 Liquid metal recorder, patent no. US3596285A, United States

[7] Singh A and Singh H 2022 Metal Additive Manufacturing: From History to Applications *Innovations in Additive Manufacturing* Springer Tracts in Additive Manufacturing ed M A Khan and J T W Jappes (Cham: Springer International Publishing) pp 3–32

[8] B.a P, N L, Buradi A, N S, B l P and R V 2022 A comprehensive review of emerging additive manufacturing (3D printing technology): Methods, materials, applications, challenges, trends and future potential *Mater. Today Proc.* **52** 1309–13

[9] Hashmi A W, Mali H S and Meena A 2022 A comprehensive review on surface quality improvement methods for additively manufactured parts *Rapid Prototyp. J.* **29** 504–57

[10] Gunpinar E and Cam S 2022 4 and 5-Axis additive manufacturing of parts represented using free-form 3D curves *Graph. Models* **120** 101137

[11] Korkmaz M E, Waqar S, Garcia-Collado A, Gupta M K and Krolczyk G M 2022 A technical overview of metallic parts in hybrid additive manufacturing industry *J. Mater. Res. Technol.* **18** 384–95

[12] Kwon S and Oh Y 2023 Optimal process planning for hybrid additive–subtractive manufacturing using recursive volume decomposition with decision criteria *J. Manuf. Syst.* **71** 360–76

[13] Abe T, Kataoka T, Kayashima S, Sumi N, Hashimoto T and Kaneko J 2023 Determination of alternation sequence for additive and subtractive manufacturing based on subtractive manufacturing simulation *Precis. Eng.* **80** 160–70

[14] Borgianni Y, Pradel P, Berni A, Obi M and Bibb R 2022 An investigation into the current state of education in Design for Additive Manufacturing *J. Eng. Des.* **33** 461–90

[15] Ranjan R, Kumar D, Kundu M and Chandra Moi S 2022 A critical review on Classification of materials used in 3D printing process *Mater. Today Proc.* **61** 43–9

[16] Dadkhah M, Mosallanejad M H, Iuliano L and Saboori A 2021 A Comprehensive Overview on the Latest Progress in the Additive Manufacturing of Metal Matrix Composites: Potential, Challenges, and Feasible Solutions *Acta Metall. Sin. Engl. Lett.* **34** 1173–200

[17] Blakey-Milner B, Gradl P, Snedden G, Brooks M, Pitot J, Lopez E, Leary M, Berto F and du Plessis A 2021 Metal additive manufacturing in aerospace: A review *Mater. Des.* **209** 110008

[18] Fé-Perdomo I L, Ramos-Grez J A, Beruvides G and Mujica R A 2021 Selective laser melting: lessons from medical devices industry and other applications *Rapid Prototyp. J.* **27** 1801–30

[19] Casini M 2022 Chapter 8 - Advanced building construction methods *Construction 4.0* Woodhead Publishing Series in Civil and Structural Engineering ed M Casini (Woodhead Publishing) pp 405–70

[20] Zhang B, Li Y and Bai Q 2017 Defect Formation Mechanisms in Selective Laser Melting: A Review *Chin. J. Mech. Eng.* **30** 515–27

[21] Ribeiro K S B, Mariani F E and Coelho R T 2020 A Study of Different Deposition Strategies in Direct Energy Deposition (DED) Processes *Procedia Manuf.* **48** 663–70

[22] Lalegani Dezaki M, Serjouei A, Zolfagharian A, Fotouhi M, Moradi M, Ariffin M K A and Bodaghi M 2022 A review on additive/subtractive hybrid manufacturing of directed energy deposition (DED) process *Adv. Powder Mater.* **1** 100054

[23] Li S-H, Kumar P, Chandra S and Ramamurty U 2023 Directed energy deposition of metals: processing, microstructures, and mechanical properties *Int. Mater. Rev.* **68** 605–47

[24] Dass A and Moridi A 2019 State of the Art in Directed Energy Deposition: From Additive Manufacturing to Materials Design *Coatings* **9** 418

[25] Li M, Du W, Elwany A, Pei Z and Ma C 2020 Metal Binder Jetting Additive Manufacturing: A Literature Review *J. Manuf. Sci. Eng.* **142**

[26] Mirzababaei S and Pasebani S 2019 A Review on Binder Jet Additive Manufacturing of 316L Stainless Steel *J. Manuf. Mater. Process.* **3** 82

[27] Lores A, Azurmendi N, Agote I and Zuza E 2019 A review on recent developments in binder jetting metal additive manufacturing: materials and process characteristics *Powder Metall.* **62** 267–96

[28] Ziaee M and Crane N B 2019 Binder jetting: A review of process, materials, and methods *Addit. Manuf.* **28** 781–801

[29] Aboulkhair N T, Bosio F, Gilani N, Phutela C, Hague R J M and Tuck C J 2023 Chapter Six - Additive manufacturing processes for metals *Quality Analysis of Additively Manufactured Metals* ed J Kadkhodapour, S Schmauder and F Sajadi (Elsevier) pp 201–58

[30] Zhang Y, Jarosinski W, Jung Y-G and Zhang J 2018 2 - Additive manufacturing processes and equipment *Additive Manufacturing* ed J Zhang and Y-G Jung (Butterworth-Heinemann) pp 39–51

[31] Khan A, Singh A K and Dugala N S 2024 Leverage of Metal 3D Printing Technology in the Automotive Industry *Machining and Additive Manufacturing* Lecture Notes in Mechanical Engineering ed V S Sharma, U S Dixit, A Gupta, R Verma and V Sharma (Singapore: Springer Nature) pp 159–72

[32] Zhang Y, Wu L, Guo X, Kane S, Deng Y, Jung Y-G, Lee J-H and Zhang J 2018 Additive Manufacturing of Metallic Materials: A Review *J. Mater. Eng. Perform.* **27** 1–13

[33] Haddad M, Nixon K B and Wolff S 2023 Sheet Lamination *Springer Handbook of Additive Manufacturing* Springer Handbooks ed E Pei, A Bernard, D Gu, C Klahn, M Monzón, M Petersen and T Sun (Cham: Springer International Publishing) pp 407–23

[34] Bhuvanesh Kumar M and Sathiya P 2021 Methods and materials for additive manufacturing: A critical review on advancements and challenges *Thin-Walled Struct.* **159** 107228

[35] Milton S, Rigo O, LeCorre S, Morandeau A, Siriki R, Bocher P and Leroy R 2021 Microstructure effects on the machinability behaviour of Ti6Al4V produced by Selective Laser Melting and Electron Beam Melting process *Mater. Sci. Eng. A* **823** 141773

[36] Bartlett J L and Li X 2019 An overview of residual stresses in metal powder bed fusion *Addit. Manuf.* **27** 131–49

[37] Ponnusamy P, Rahman Rashid R A, Masood S H, Ruan D and Palanisamy S 2020 Mechanical Properties of SLM-Printed Aluminium Alloys: A Review *Materials* **13** 4301

IOP Conf. Series: Materials Science and Engineering **1313** (2024) 012013
doi:10.1088/1757-899X/1313/1/012013

[38] Ahmadi M, Tabary S A A B, Rahmatabadi D, Ebrahimi M S, Abrinia K and Hashemi R 2022 Review of selective laser melting of magnesium alloys: advantages, microstructure and mechanical characterizations, defects, challenges, and applications *J. Mater. Res. Technol.* **19** 1537–62

[39] Sola A and Nouri A 2019 Microstructural porosity in additive manufacturing: The formation and detection of pores in metal parts fabricated by powder bed fusion *J Adv Manuf Process* **1:e10021**

[40] Winkler K, Ahmad H, Danzig A, Gloetter P, Schubert F, Wagner G and Kroll L 2023 Effect of organic impurity in additive manufactured Ti-6Al-4V *Addit. Manuf. Lett.* **5** 100116

[41] Chahal V and Taylor R M 2020 A review of geometric sensitivities in laser metal 3D printing *Virtual Phys. Prototyp.* **15** 227–41

[42] Zhang X, Yocom C J, Mao B and Liao Y 2019 Microstructure evolution during selective laser melting of metallic materials: A review *J. Laser Appl.* **31** 031201

[43] Tucho W M, Lysne V H, Austbø H, Sjolyst-Kverneland A and Hansen V 2018 Investigation of effects of process parameters on microstructure and hardness of SLM manufactured SS316L *J. Alloys Compd.* **740** 910–25

[44] Lu H Z, Ma H W, Cai W S, Luo X, Qu S G, Wang J, Lupoi R, Yin S and Yang C 2022 Altered phase transformation behaviors and enhanced bending shape memory property of NiTi shape memory alloy via selective laser melting *J. Mater. Process. Technol.* **303** 117546

[45] Zhang Z, Chu B, Wang L and Lu Z 2019 Comprehensive effects of placement orientation and scanning angle on mechanical properties and behavior of 316L stainless steel based on the selective laser melting process *J. Alloys Compd.* **791** 166–75

[46] Nath S D, Gupta G, Kearns M, Gulsoy O and Atre S V 2020 Effects of layer thickness in laser-powder bed fusion of 420 stainless steel *Rapid Prototyp. J.* **26** 1197–208

[47] Ni C, Zhu L, Zheng Z, Zhang J, Yang Y, Hong R, Bai Y, Lu W F and Wang H 2020 Effects of machining surface and laser beam scanning strategy on machinability of selective laser melted Ti6Al4V alloy in milling *Mater. Des.* **194** 108880

[48] Al-Rubaie K S, Melotti S, Rabelo A, Paiva J M, Elbestawi M A and Veldhuis S C 2020 Machinability of SLM-produced Ti6Al4V titanium alloy parts *J. Manuf. Process.* **57** 768–86

[49] Wu X, Li L, He N, Yao C and Zhao M 2016 Influence of the cutting edge radius and the material grain size on the cutting force in micro cutting *Precis. Eng.* **45** 359–64

[50] Zheng Z, Jin X, Bai Y, Yang Y, Ni C, Lu W F and Wang H 2022 Microstructure and anisotropic mechanical properties of selective laser melted Ti6Al4V alloy under different scanning strategies *Mater. Sci. Eng. A* **831** 142236

[51] Le Coz G, Fischer M, Piquard R, D'Acunto A, Laheurte P and Dudzinski D 2017 Micro Cutting of Ti-6Al-4V Parts Produced by SLM Process *Procedia CIRP* **58** 228–32

[52] Komatsu T, Yoshino T, Matsumura T and Torizuka S 2012 Effect of Crystal Grain Size in Stainless Steel on Cutting Process in Micromilling *Procedia CIRP* **1** 150–5

[53] Polishetty A, Nomani J and Littlefair G 2023 Evaluating and comparing secondary machining characteristics of wrought and additive manufactured 316L stainless steel *Mater. Today Proc.*

[54] Shunmugavel M, Polishetty A, Goldberg M, Singh R and Littlefair G 2017 A comparative study of mechanical properties and machinability of wrought and additive manufactured (selective laser melting) titanium alloy – Ti-6Al-4V *Rapid Prototyp. J.* **23** 1051–6

Acknowledgements
This research was partly supported by the National Research, Development and Innovation Office through the projects ED_18-2018-0006 (Research on prime exploitation of the potential provided by the industrial digitalisation), OTKA-K-132430 (Transient deformation, thermal and tribological processes at fine machining of metal surfaces of high hardness), and by the European Commission through the H2020 project EPIC under grant No. 739592

IOP Conf. Series: Materials Science and Engineering **1313** (2024) 012014 doi:10.1088/1757-899X/1313/1/012014

Manufacturing of composites with designed failure

G Zs Marton[1,2], Á Fendrik[3] and G Szebényi[1,2]

[1] Department of Polymer Engineering, Faculty of Mechanical Engineering, Budapest University of Technology and Economics, Műegyetem rkp. 3., H-1111 Budapest, Hungary

[2] MTA-BME Lendület Sustainable Polymers Research Group, Műegyetem rkp. 3., H-1111 Budapest, Hungary

[3] MouldTech Systems Kft., Fuvar u. 7., H-8900 Zalaegerszeg

E-mail: szebenyi@pt.bme.hu

Abstract: The failure of polymer composites is a very complex process, which mainly occurs as a result of interactions between different types of damage in the material, often at random locations and without any particular signs. To increase the reliability of composites and enable their wider use in safety-critical components, it is crucial to make their failure processes more controllable and predictable, which could be achieved by the local modification of the interfacial adhesion. The aim of our research is to develop and investigate a method for designing the failure process of polymer composites in terms of the location and the mode of failure. We produced interfacially engineered composites containing a weakened adhesion zone by adding polycaprolactone (PCL) thermoplastic interlayer material, which was applied directly on the surface of the reinforcing material by fused filament fabrication (FFF). Then, material tests were carried out, and the failure mode and position were evaluated as a function of the geometry of the interlayer material.

1. Introduction

Nowadays, polymer composites are becoming more and more common, mainly due to their excellent specific mechanical properties – especially strength and stiffness. Besides mass products, their applications include structural components as well. However, their further spread may be limited by their failure process, which can be considered unfavourable. Their heterogeneous microstructure, the significant differences in the properties of the reinforcement material and matrix, the presence of interfaces, and the anisotropic properties of composites all contribute to the complexity of their damage and failure mechanisms [1, 2].

Polymer composites contain different types of damage in a large number. Defects can be divided into several categories. Based on the source of the damage, material defects, manufacturing-induced flaws, and in-service damage can be distinguished. Damage can also vary in size, varying from the nano to the macro scale. The four basic types of damage are fiber failure – breakage or micro-buckling – matrix cracking, delamination, and fiber-matrix debonding [2-4].

Different forms of damage can occur simultaneously in the material under load. Their evolution and interactions can lead to the failure of the structure, which can take place in random positions or in several locations at the same time, often catastrophically [2-4].

Avoiding catastrophic failure by increasing the toughness of composites and achieving pseudo-ductile behaviour is a hot topic. There are several solutions for this purpose. Interface modification provides a great opportunity for it, for example, by applying nanofiber mats or polymer films [5, 6]. The application of polycaprolactone (PCL) interlayer material is also a promising interface engineering technique, which enables the local modification of adhesion. It is important that PCL is soluble in the epoxy matrix and does not create a new phase. The application of PCL interlayer material seems successful in modifying the failure behaviour and increasing the ductility of the material [7].

A further disadvantage of polymer composites is that their damages are challenging to detect and localise, but non-destructive testing (NDT) methods can offer a solution. Alongside widespread NDT methods such as acoustic emission testing and ultrasonic testing, the digital image correlation (DIC) technique is gaining popularity in the composite industry. DIC is a non-contact, optical method that has been used for a relatively long time in the materials testing of composites to determine the displacements and strain fields under load, which is able to detect the presence and propagation of meso- and macro-scale damage in the composite, based on high local strain values and changes in the strain-field [4, 8].

In our research, we manufactured composite samples containing PCL interlayer material as a locally weakened adhesion zone. The aim was to predetermine the failure in terms of location and mode. The failure process was investigated under different loading conditions.

2. Experimental

2.1. Materials

IPOX MR 3010 (IPOX Chemicals Kft., Budapest, Hungary) epoxy resin (EP) with IPOX MH 3124 curing agent was applied as matrix material. The components were mixed in a weight ratio of 100:35. We used PX35FBUD0300 (Zoltek Zrt., Nyergesújfalu, Hungary) stitch-bonded unidirectional carbon fabric (309 g/m2 surface weight), consisting of Panex35 50k roving, as reinforcement material. We applied eMorph175N05 (Shenzhen Esun Industrial Co. Ltd., Shenzhen, China) poly (ϵ-caprolactone) (PCL) filament (1.75 mm diameter, 60°C melting temperature, 180°C printing temperature) as interlayer material.

2.2. Test methods

Firstly, the fiber volume fractions of the composite plates were estimated from density measurements carried out with a Sartorius Quintix 125D (Sartorius Lab Instruments GmbH, Göttingen, Germany) weighing scale, based on the rule of mixtures.

Then, Charpy impact tests were carried out according to MSZ EN ISO 179-2 with a Ceast Fractovis 9350 impact tester equipped with a DAS 8000 data acquisition unit on composite specimens with 80 mm length and 10 mm width. The test was performed with a flatwise impact with 15 J impact energy and an impact velocity of 3.7 m/s with a support distance of 62 mm.

The tensile tests were carried out according to MSZ EN ISO 527-4 with a Zwick Z250 (Zwick GmbH, Ulm, Germany) universal testing machine equipped with a 250 kN load cell. Specimens with bonded end-tabs, a free length of 150 mm and a width of 15 mm were used for the tensile tests. The applied test speed was 5 mm/min. The displacements were measured with a Mercury Monet (Sobriety, Kurim, Czech Republic) optical strain measuring system (DIC) with a 5 MPixel resolution IDS U3-3080CP-P-GL (Imaging Development Systems Inc., Obersulm, Germany) camera and two LED lights. The results were recorded and processed with the Mercury RT-v2.9 software. For the data acquisition, a frame rate of 10 Hz was used. Before the test, the optical strain measurement system was calibrated to avoid lens distortion and to correlate the real size of the specimen with its size in the recorded images. For the global strain measurement, a gauge length of 75 mm was used with a line probe in the software. Two additional line probes were used, one for the further measurement of global strain with a length of 50 mm and one for local strain measurement in the region of the PCL interlayer material with a length of 15 mm. A strain field map was also generated by DIC, where the calculated points were 20 pixels from each other, and the facet window size was 40 pixels × 40 pixels.

2.3. Sample production

Reference composites without PCL and composite plates containing PCL interlayer material were manufactured with $[0°]_6$ layup sequence. For the latter, the PCL filament was printed directly on the surface of the carbon fabric by fused filament fabrication (FFF) method using a CraftBot Plus 3D printer. The samples manufactured for the impact tests contained interlayer material inclined at an angle of 45° to the fiber direction, so that the effect of the weakened adhesion zone on the failure location could be better observed. The interlayer material in the tensile test specimens was applied perpendicular to the fiber direction (Figure 1 and Figure 2). The interlayer material, thus the weakened adhesion zone, differed in geometry: the zone had the same width for each plate, only its thickness varied. Composite plates were produced by vacuum-assisted resin transfer moulding (VARTM). Then, specimens were cut from the composite plates with a Mutronic Diadisc (Mutronic, Rieden am Forggensee, Germany) diamond disc cutter.

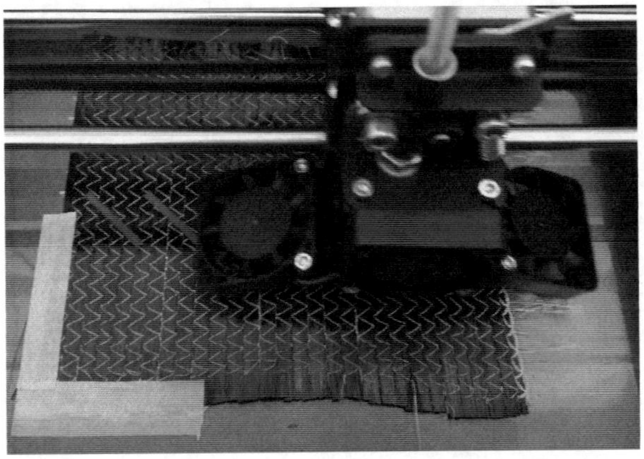

Figure 1. Printing of PCL on the UD carbon fabric by FFF method for the impact test specimen.

Figure 2. UD carbon fabric containing the PCL interlayer material during the manufacturing of the tensile test specimen.

3. Results

The average fiber volume fraction of the composite plates was 41.6±1.7 V/V%. These differences do not significantly affect the mechanical properties and the material behaviour.

3.1. Results of the impact tests

During the impact tests, the following samples were investigated:

- Reference samples containing no PCL,
- Specimens containing PCL interlayer material with a width of 2 mm and a thickness of 0.2 mm,
- Specimens containing PCL interlayer material with a width of 2 mm and a thickness of 0.8 mm.

First, the dynamic mechanical properties of the different composites were determined from the measured data (Table 1).

Table 1. The dynamic mechanical properties of the different composites.

	Impact strength [kJ/m^2]	Ductility index [-]
Reference	124.5±13.6	5.1±1.6
PCL with 2 mm width, 0.2 mm thickness	93.8±8.7	5.4±1.6
PCL with 2 mm width, 0.8 mm thickness	88.0±6.8	48.9±9.0

Table 1 shows that the ductility index is only significantly affected by the application of the 0.8 mm thick, 2 mm wide interlayer material, which results in a significant increase compared to the reference samples. In addition, it is important to note that the impact strength shows a decrement in case of composites containing interlayer material compared to the reference specimens, which can be explained by the presence of PCL.

Then, the failure mode and position were evaluated. The results show that the reference composites failed in a brittle manner along the line of the impact, as expected (Figure 3). For the composites containing PCL interlayer material, Figure 4 and Figure 5 suggest that the smaller thickness does not significantly affect either the fracture path or the mode of failure. However, the 0.8 mm thick interlayer material had a significant effect on both failure mode and position: two specimens failed by delamination, while for three specimens, the fracture path was increased, with failure occurring along the line of the interlayer material.

Figure 3. Reference specimens after the impact test.

Figure 4. The specimens containing PCL interlayer material with a width of 2 mm and a thickness of 0.2 mm after the impact test.

Figure 5. The specimens containing PCL interlayer material with a width of 2 mm and a thickness of 0.8 mm after the impact test.

Impact tests suggest that it is possible to influence the mode and position of failure by using PCL interlayer material, thus creating zones of weakened adhesion.

3.2. Results of the tensile tests

In the tensile tests, we focused more on the effect of the geometry of the PCL interlayer material, so we investigated the following composites:

- Reference samples containing no PCL,
- Specimens containing PCL interlayer material with a width of 2 mm and a thickness of 0.2 mm,
- Specimens containing PCL interlayer material with a width of 2 mm and a thickness of 0.4 mm,

- Specimens containing PCL interlayer material with a width of 2 mm and a thickness of 0.8 mm.

First, the mechanical properties – tensile strength, tensile modulus and strain at break – of the different composite plates were determined (Table 2 and Figures 6-8).

Table 2. The tensile properties of the different composites.

	Tensile strength [MPa]	Tensile modulus [GPa]	Strain at break [%]
Reference	993.7±78.3	98.7±2.5	1.15±0.12
PCL with 2 mm width, 0.2 mm thickness	1039.0±67.0	73.7±3.3	1.26±0.19
PCL with 2 mm width, 0.4 mm thickness	968.5±60.6	71.3±12.2	1.37±0.15
PCL with 2 mm width, 0.8 mm thickness	994.6±71.9	73.1±2.8	1.16±0.09

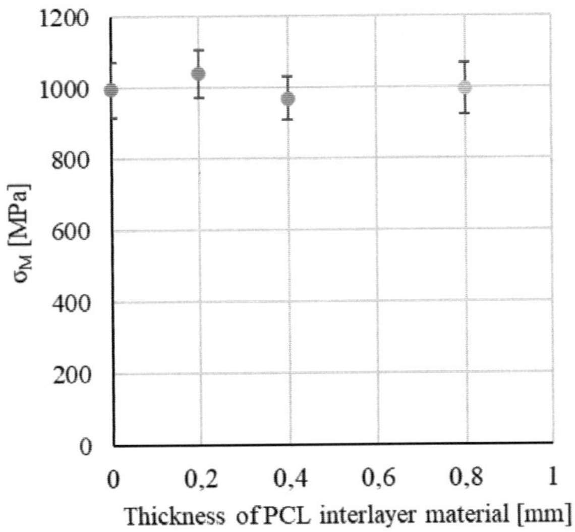

Figure 6. The effect of the thickness of PCL interlayer material on the tensile strength

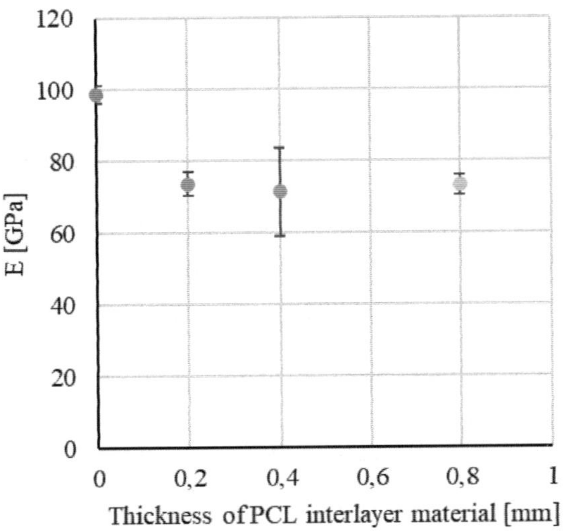

Figure 7. The effect of the thickness of PCL interlayer material on the tensile modulus

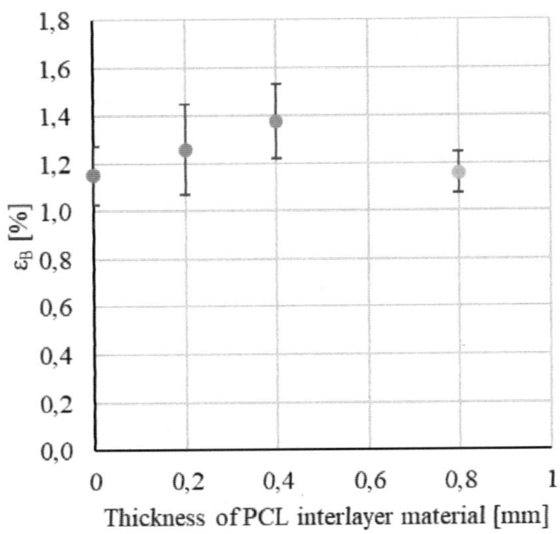

Figure 8. The effect of the thickness of PCL interlayer material on the strain at break

The results suggest that using interlayer material only leads to a significant change in the tensile modulus. The value of the tensile modulus decreases with the application of the interlayer material. However, the measured data show that for the same width, the thickness of the interlayer material does not have a significant effect on the value of the modulus.

Next, data from Line Probe (LP) and full-field measurements, recorded by DIC during the tensile tests were evaluated, which allowed the detection of possible damage, especially delamination. During the line probe test, we determined the elongation along three lines with different lengths, in the arrangement shown in Figure 9, so that line probes A2 and A3 provided information on the global strain, while A1 provided information on the local strain in the region of the interlayer material. The test was also performed for the reference specimens. The results of the line probe tests can be seen in Figures 10-13 for one representative specimen per specimen type.

Figure 9. Arrangement of the line probe test, A1= 15 mm for local strain measurement, A2=50 mm and A3=75 mm for global strain measurement

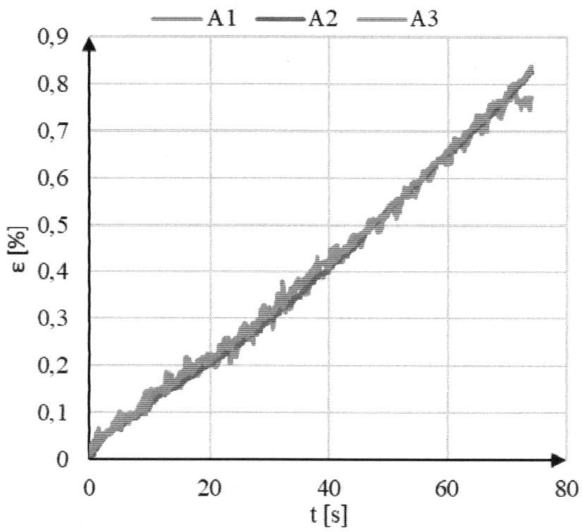

Figure 10. Typical result of a line probe test of a reference specimen

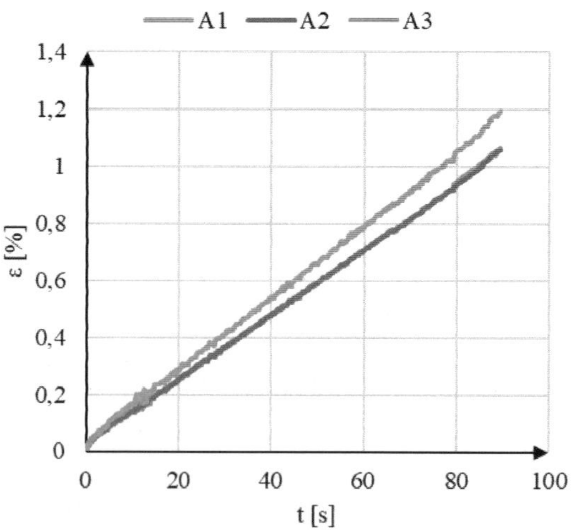

Figure 11. Typical result of a line probe test of a specimen with 2 mm wide, 0.2 mm thick PCL

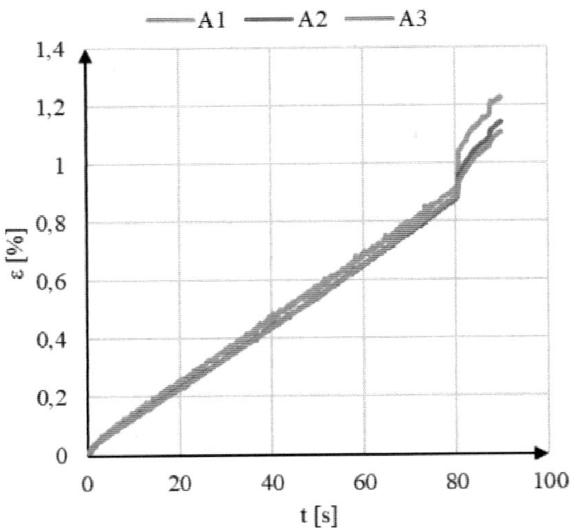

Figure 12. Typical result of a line probe test of a specimen with 2 mm wide, 0.4 mm thick PCL

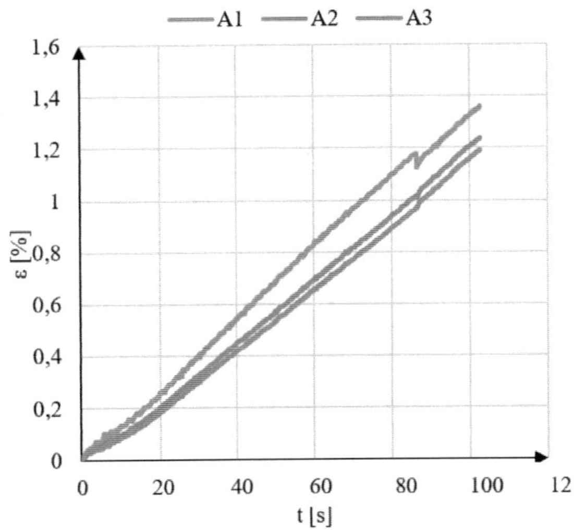

Figure 13. Typical result of a line probe test of a specimen with 2 mm wide, 0.8 mm thick PCL

The results of the line probe tests show that the local strain exceeds the global strain for specimens containing interlayer material, indicating the failure location and the presence of possible local delamination, likely as a result of the weakened adhesion zone. In contrast, there is no significant difference in the strains measured by the line probes for the reference sample.

We determined the strain-fields (50 mm x 15 mm) at the same global strain value for each specimen type (Figures 14-17).

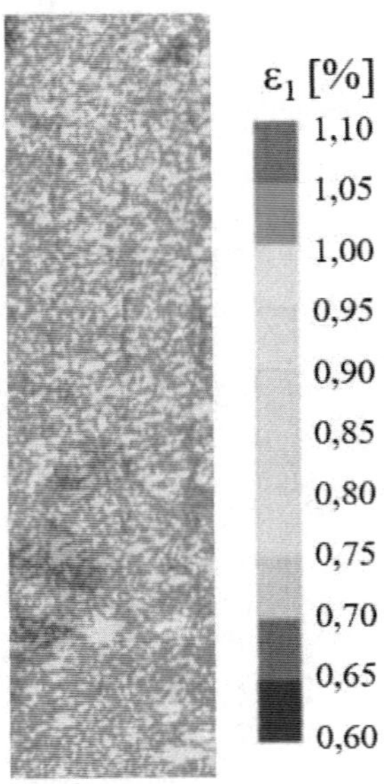

Figure 14. Typical strain field of a reference specimen

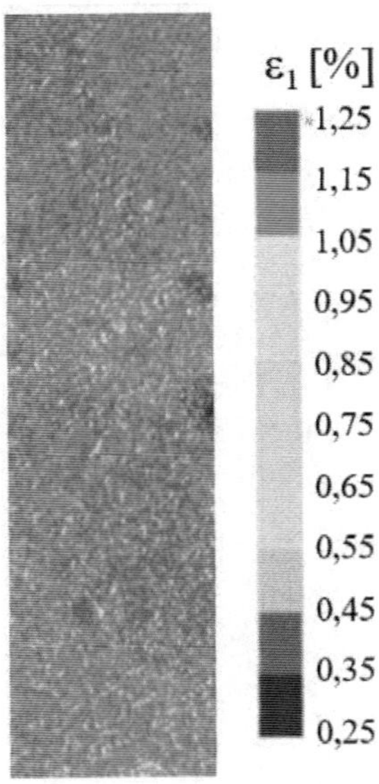

Figure 15. Typical strain field of a specimen with 2 mm wide, 0.2 mm thick PCL

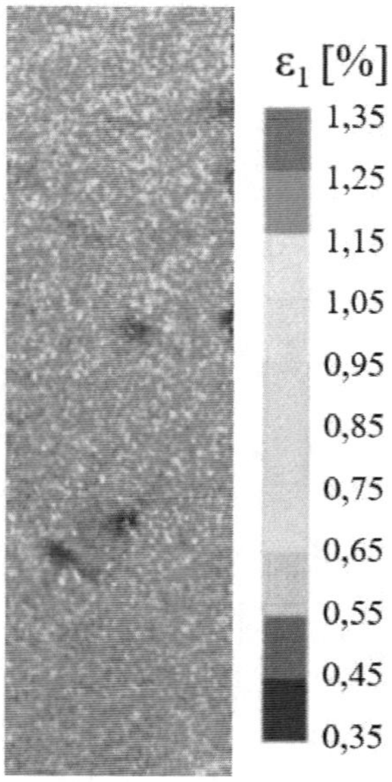

Figure 16. Typical strain field of a specimen with 2 mm wide, 0.4 mm thick PCL

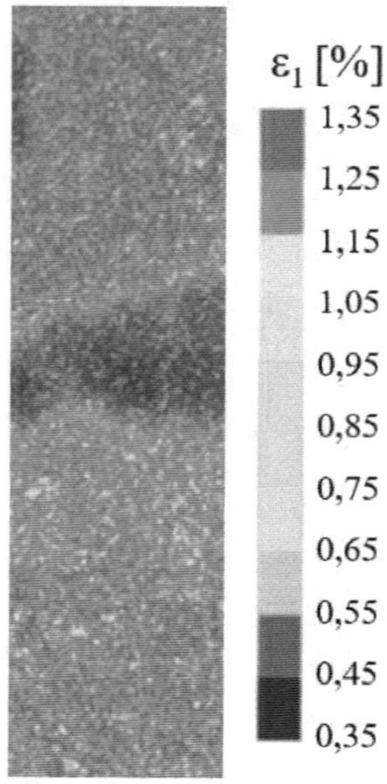

Figure 17. Typical strain field of a specimen with 2 mm wide, 0.8 mm thick PCL

The results of the full-field strain measurements confirmed the conclusions of the line probe test. The strain field of the reference sample is quite homogenous. It only contains a random failure, while the weakened adhesion zones formed by the PCL interlayer material cause a local increase in strain, predicting the location of the failure and indicating the formation of a delamination in the region. The interlayer material with 0.8 mm thickness has the most significant effect, which, in addition to modifying the adhesion, might create higher local strain and possible delamination through the formation of local fiber waviness and resin-rich areas.

4. Conclusion

We manufactured composite plates containing PCL interlayer material for the local modification of interlaminar adhesion. We investigated the effect of the application of PCL and the geometry of the weakened adhesion zone on the failure process and the mechanical properties of composite plates. We found that the use of PCL interlayer material enables the design of the failure mode and position, and the effectiveness of the method depends on the geometry of the weakened adhesion zone. The tensile tests suggest that the mechanical properties do not decrease significantly, while the impact test indicates a decrement in the impact strength but an increment in the ductility index.

5. References

[1] Talreja R and Singh C V 2012 *Damage and Failure of Composite Materials* (Cambridge: Cambridge University Press)

[2] Heslehurst R B 2014 *Defects and Damage in Composite Materials and Structures* (Boca Raton: CRC Press)

[3] Wang B, Zhong S C, Lee T L, Fancey K S and Mi J W 2020 Non-destructive testing and evaluation of composite materials/structures: A state-of-the-art review *Advances in Mechanical Engineering.* **12** 1-28

[4] Duchene P, Chaki S, Ayadi A and Krawczak P 2018 A review of non-destructive techniques used for mechanical damage assessment in polymer composites *Journal of Materials Science.* **53** 7915-7938

[5] Molnar K, Kostakova E and Meszaros L The effect of needleless electrospun nanofibrous interleaves on mechanical properties of carbon fabrics/epoxy laminates *Express Polymer Letters.* **8** 62-72
[6] Marino S G and Czél G 2021 Improving the performance of pseudo-ductile hybrid composites by film-interleaving *Composites Part A.* **142** 106233/1-106233/16
[7] Magyar B, Czigány T and Szebényi G 2021 Metal-alike polymer composites: The effect of inter-layer content on the pseudo-ductile behaviour of carbon fibre/epoxy resin materials *Composites Science and Technology.* **215** 109002/1-109002/8
[8] Hliva V and Szebényi G 2023 Non-Destructive Evaluation and Damage Determination of Fiber-Reinforced Composites by Digital Image Correlation *Journal of Nondestructive Evaluation.* **42** 43/1-43/15

Acknowledgments
The research has been supported by the NRDI Office (OTKA FK 142540, 2019-1.1.1-PIACI-KFI-2019-00139) Project no. 2022-2.1.1-NL-2022-00012 „National Laboratory for Cooperative Technologies" has been implemented with the support provided by the Ministry of Culture and Innovation of Hungary from the National Research, Development and Innovation Fund, financed under the National Laboratories funding scheme. Gábor Szebényi acknowledges the financial support received through János Bolyai Scholarship of the Hungarian Academy of Sciences and through the ÚNKP-23-5-BME-415 New National Excellence Program of the Ministry for Innovation and Technology. The project supported by the Doctoral Excellence Fellowship Programme (DCEP) is funded by the National Research Development and Innovation Fund of the Ministry of Culture and Innovation and the Budapest University of Technology and Economics, under a grant agreement with the National Research, Development and Innovation Office.

The effect of building strategies on tensile strength of additively manufactured parts by laser metal deposition

G Juhász[1,2,3], **F Hareancz**[1,2,3], **G Buza**[1], **R E Fábián**[3] and **Á Vida**[1]

[1] Bay Zoltán Nonprofit Ltd. for Applied Research, Kondorfa Street 1, Budapest, 1116, Hungary

[2] Department of Innovative Vehicles and Materials, GAMF Faculty of Engineering and Computer Science, John Von Neumann University, Izsáki Street 10., Kecskemét, 6000, Hungary

[3] Doctoral School on Materials Sciences and Technologies, Óbuda University, Népszínház Street 8., Budapest 1081, Hungary

E-mail: Adam.Vida@bayzoltan.hu

Abstract. Laser Metal Deposition (LMD) is a powder-based technology. This technology can be used for surface coating, repairing and additive manufacturing. The property of the part is dependent on the building parameters of the process. In the research work, we use austenitic steel (316L) powder to produce blocks with different building strategies, from which we machine tensile test specimens. The patterns were prepared according to three strategies: the welded bands were parallel to the longitudinal axis of the test specimen, perpendicular to it, and finally the parallel and perpendicular directions were alternated. We perform hardness and tensile tests on these specimens and search for correlations between the strength and the strategies.

1. Introduction

Laser technology is growing in research and industry. The technology can be used to modify various surfaces with or without a filler material [1, 2, 3]. Laser cladding can be used to modify a surface with a filler material [4], or even to build a body. This technology uses wire or metal powder. The use of metal powders is a growing area of industrial manufacturing. Examples include powder metallurgy (PM) [5], additive manufacturing (AM) [6,7], and metal injection moulding (MIM) [8]. Metal powder-based additive manufacturing generally uses lasers as a heat source. This technology has a wide range of industrial applications, including rapid prototyping, component repair, surface coating and innovative alloy development. Typically, this technology can be classified into two processes: powder bed and powder deposition. Selective Laser Sintering (SLS) and Selective Laser Melting (SLM) are laser-based powder bed technologies that can produce fine, accurate, and complex parts. However, these parts can only be made from one type of powder, and their size are limited. Parts produced using powder deposition techniques such as Laser Metal Deposition (LMD), Direct Metal Deposition (DMD) or Direct Energy Deposition (DED) are stronger but can be less complex and have rougher surfaces. This technique also allows for surface coating and body repair, which is not possible with powder bed technology [9, 10]. This is an additional positive feature of this technique.

In the LMD process, the first step is to heat the base plate. A melted area forms on the surface of the baseplate, and then metal powder is blown into the melt pool by inert gas, such as argon. In the case of a coaxial LMD head, the powder is blown in the optical path of the laser, which heats the powder. The metal powder is not melted until it enters the melt pool. This process results in a layer with an altered chemical composition along the laser path relative to the workpiece. If necessary, several layers can be built on top of each other to create bodies. Figure 1 [11] shows a schematic of the process.

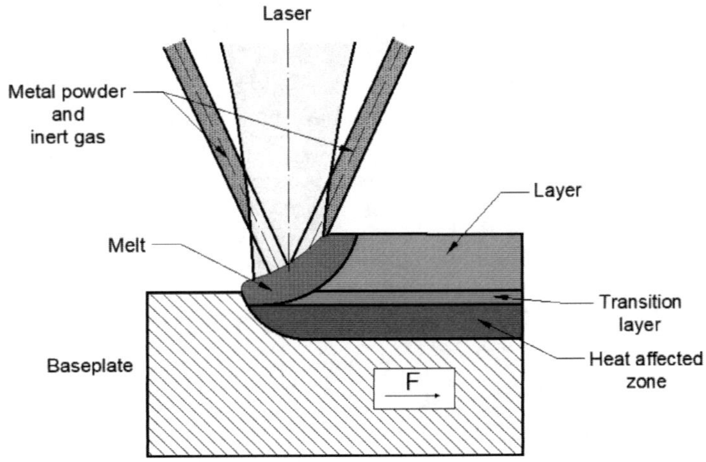

Figure 1. Schematic figure of the Laser Metal Deposition process [11].

When 3D parts are created using this technique, the built body undergoes several heating, melting, and cooling cycles, which develop complex microstructural structures. It can be concluded that the process strategy influences the properties of the finished part [12].
The purpose of this work is to determine the effect of different building strategies on the mechanical properties of an LMD-created test block. To achieve this, we machined tensile test specimens from the blocks. Additional blocks were created for hardness testing, and the microstructure of the specimens was investigated.

2. Experimental section

2.1. Materials
For the LMD process we used a Oerlikon Metco 41C stainless steel metal powder. The chemical composition of the metal powder is shown in Table 1.

Table 1. Chemical composition of the testing material

Material	Fe%	C%	Mo%	Ni%	P%	S%	Si%
Metco 41 C (316L)	66.15	0.03	2.5	12	0.01	0.01	2.3

The powder is water atomized resulting in irregularly shaped particles with a diameter ranging from 45 to 150 μm. The baseplate substrate is made of a stainless steel plate with dimensions of 150 mm x 100 mm x 12 mm. Prior to the process, the substrate was cleaned.

2.2. Laser equipment and parameters
Experiments were carried out using a Trumpf 7020 cell equipped with LMD technology and a 4 kW solid-state laser source. The machining optics consisted of a Trumpf MSO70 LMD head with a SO12 nozzle. The powder feeder used was a GTV PF/2. The nozzle was positioned 12 mm from the baseplate, which is the focal length of the powder stream. The laser focus diameter was 3 mm. We determined the optimal laser parameters through preliminary research, which are summarised in Table 2.

Table 2. Laser parameters of the LMD process

Power [W]	1250
Feed rate [m/min]	0.6
Powder feed rate [g/min]	6.5
Shielding gas flow [l/min]	8
Hatch offset [mm]	0.9
Distance between each layer [mm]	0.9

On each base plate we made a 65 mm x 20 mm x 15 mm block with these LMD parameters with different strategies. There were three building strategies, for the first type the welding beads are parallel with the loading of the tensile specimen. In the case of the second type the loading and welding directions are perpendicular with each other. The third type is bidirectional, which means one layer is like the first type, the other layer is like the second type. The bidirectional type and the location of the place where the tensile specimen is machined out are illustrated in Figure 2.

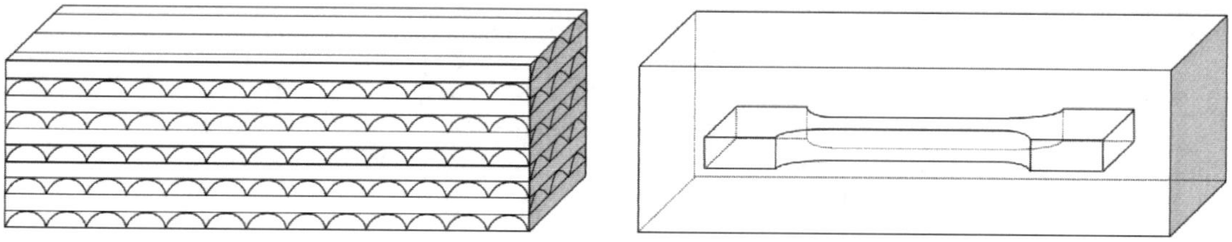

Figure 2. Schematic illustration of the bidirectional strategy and the location of the test specimen.

The dimensions of the tensile specimens are illustrated in Figure 3.

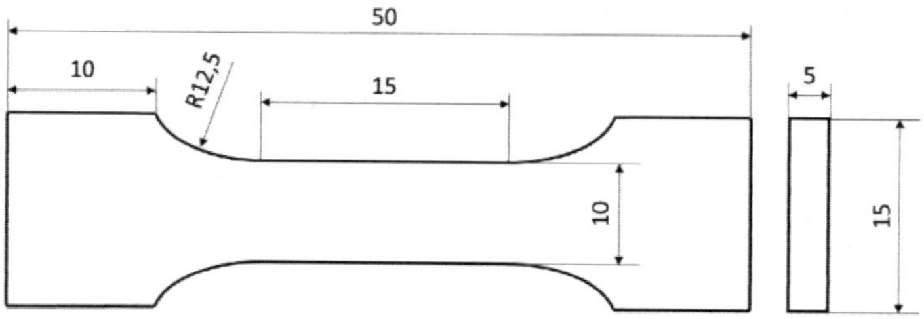

Figure 3. The dimensions of the tensile test specimen.

The specimen for tensile testing was machined using CNC milling.

2.3. Characterization

A Zeiss AxioImager 2 optical microscope was used to examine the microstructure. Separate test blocks were created and cut in two ways to investigate the layers parallel and perpendicular to the cladding direction. General metallographic techniques were used to grind and polish the specimens, followed by etching with diluted Aqua regia.

Hardness investigations were performed using a Struers Duramin 100 automatic hardness measurement machine. The hardness tests were conducted on the same specimens used in the metallographic investigation. Figure 4 shows the location of the hardness tests.

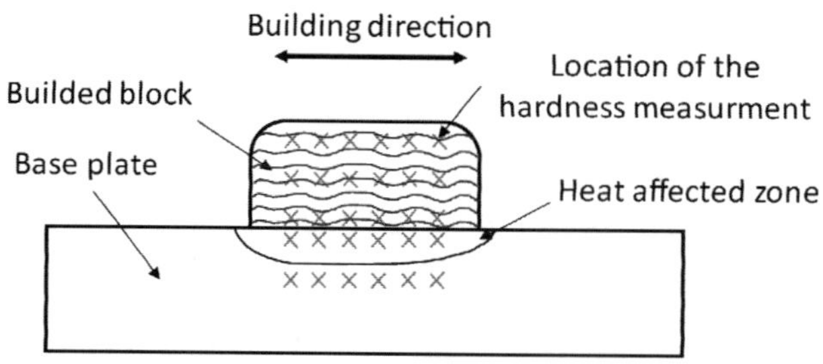

Figure 4. Schematic illustration of the hardness tests in the case of parallel to the building direction.

A universal electromechanical tensile testing machine INSTRON 3366 was used for the tensile tests. Tests were performed at room temperature, the speed of the crosshead was 5 mm/min. The resulting data was plotted in an engineering stress versus strain diagram.

3. Results and discussion

3.1. Optical microscopy results

Figure 4 displays the results of the optical microscopy. The curved shape of the layers is due to surface tension effects during the LMD process. The samples contain porosities of approximately 50 microns or less. Upon closer inspection, three types of microstructures can be identified, which are a combination of cellular, dendritic, fine, and coarse morphologies. These differences arise from varying cooling rates, which impact the formation of the resulting microstructure [8].

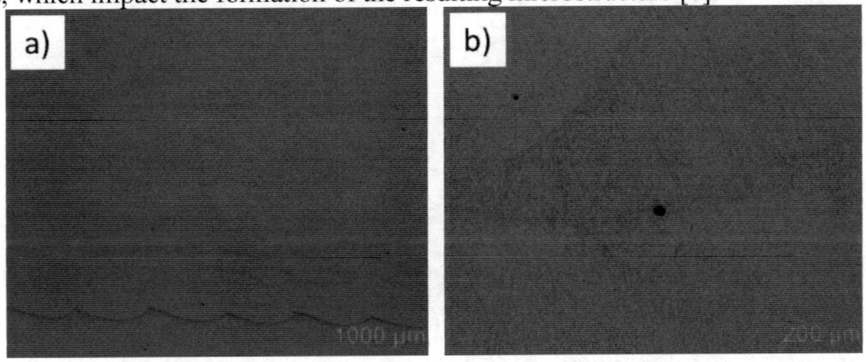

Figure 5. Optical microscopy images of the test block in a) 25x b) 100x magnification.

There are no significant differences between the building strategies in the case of microstructures.

3.2. Hardness measurement results

The base plate has a hardness of 208 HV0.5. The layer hardness measurements do not show significant differences between the strategies. The highest hardness value of the blocks is 177 HV0.5, which belongs to the first block type built parallel to the loading of the tensile specimen. The perpendicular building strategy has the lowest average hardness value, which is 167 HV0.5. The bidirectional strategy has a hardness value of 170 HV0.5, which falls between the values of the other strategies. Figure 6 displays these results.

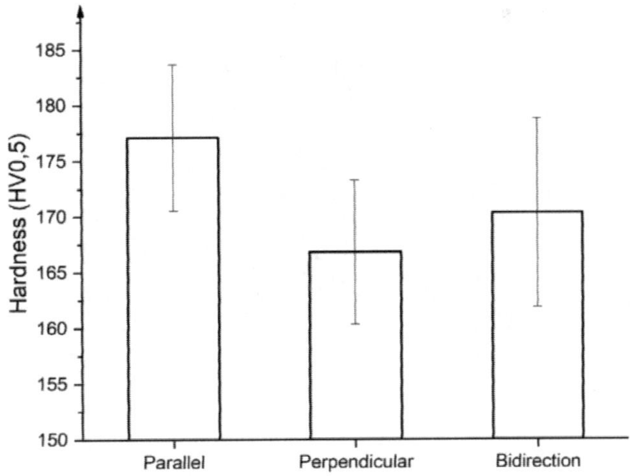

Figure 6. Hardness measurement results of layers built with different strategies.

The internal layers of the block have lower hardness than the layers closer to the base plate, as shown in Figure 7. This is due to the higher cooling rates during the building process when it is closer to the base plate.

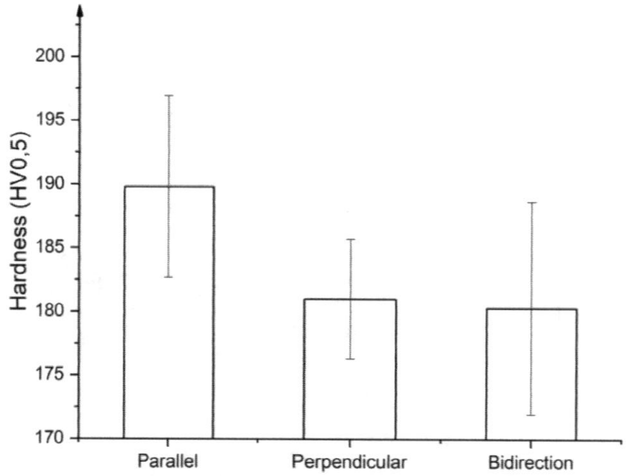

Figure 7. Hardness measurement results of layers that are close to the base plate, built with different strategies.

For layers that are closer than 1 mm to the base plate, the parallel strategy has the highest value of 190 HV0.5, while the bidirectional strategy has a similar hardness of 180 HV0.5 to the perpendicular strategy. The deviation is greater for the bidirectional strategy, which may have caused this result.

3.3. Tensile test results

The fractured surfaces of the specimens were examined by optical microscopy. The fractured surface of one of the specimens is shown in Figure 8. There are no bonding defects on the surface, the fracture resembles a ductile type of failure.

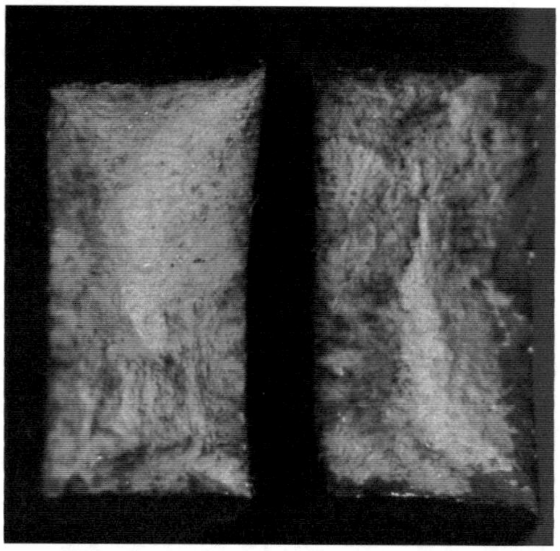

Figure 8. Fracture surface of a broken tensile sample.

Figure 9 shows the results of the tensile test. The results indicate that there is no significant difference between the differently built samples, except for the sample that was built perpendicular to the loading. The bidirectional sample has the highest tensile strength at 665 MPa, followed by the parallel sample at 646 MPa and the perpendicular sample at 617 MPa. The classical method was used to determine the $Rp_{0,2}$ yield limit. The yield limit of the parallel and bidirectional samples is 415 MPa, while the perpendicular type has a limit of 350 MPa. The inaccurate measurement of the upper crosshead displacement and the small dimensions of the tensile specimens are responsible for the high deformation values. Our results are consistent with the literature, except for the elongation value.

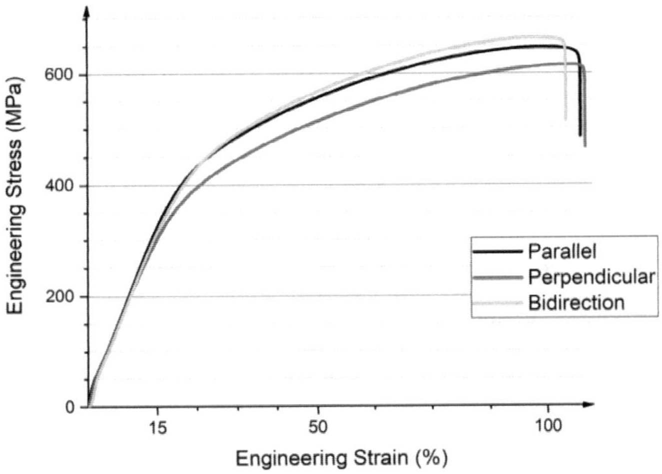

Figure 9. Tensile strength diagrams of the differently built samples.

4. Conclusion

Laser metal deposition is a suitable technology for additive manufacturing. The research involved producing test blocks using different build strategies. These blocks were used to produce a series of tensile test specimens. Blocks were also produced for hardness and metallographic testing.

The different building strategies did not result in different microstructures. The specimens had cellular, dendritic microstructures, there were areas where the particles were finer or coarser. Only 50 microns or less of porosity could be seen in the internal layers. From the hardness test results, the parallel type specimens had higher average hardness values (190 HV0.5), but in the tensile test, the bidirectional strategy gave the best tensile strength. In conclusion, by choosing the right technological setup for laser metal deposition, a homogeneous structure can be obtained which is not significantly dependent on the building strategy.

References

[1] Kovács P, Körömi B, Weltsch Z and Berczeli M 2023 *Mater.* **16**, 7663
[2] Berczeli M and Weltsch Z 2018 *IOP Conf. Ser. Mater. Sci. Eng.* **448** (2018)012050
[3] Berczeli M 2020 *Acta. Phys. Pol. A.* **137** 864-7
[4] Li J, Chen C and Wang D 2012 *Composites, Part B.* **43** 1207-12
[5] Liu C, Jiang X, Sun H, Liu TT, Wu Z and Yang L 2023 *J. Mater. Res. Technol.* **27** 7614-26
[6] Miao H, Yusof F, Karim MSA, Wu B, Raja S, Ibrahim MZ, Aziz I and Chen DL 2023 *J. Mater. Res. Technol.* **26** 8351-65
[7] Yu S, Wang J, Chen C, Wang R, Xu S, Zhao J and Ren Z. 2024 *J. Mater. Res. Technol.* **28** 1276-93
[8] Ledniczky G and Weltsch Z 2023 *Acta Mater. Trans.* **6/1** 33–7
[9] Chen Z, Sun W, Huo Y, Zhou H, Yang K and Lu J 2022 *Surf. Coat. Technol.* **448** (2022)128899
[10] Rymer L-M, Lindner T and Lampke T 2023 *Surf. Coat. Technol.* **470** (2023)129832
[11] Gong G, Ye J, Chi Y, Zhao Z, Wang Z, Xia G, Du X, Tian H, Yu H and Chen C 2021 *J. Mater. Res. Technol.* **15** 855-84
[12] Mukherjee M 2019 *Mater.* **7** (2019)100359

IOP Conf. Series: Materials Science and Engineering **1313** (2024) 012016 doi:10.1088/1757-899X/1313/1/012016

Investigation of Co-Cr alloy layer prepared by laser metal deposition

F Hareancz[1,2,3], **G Juhász** [1,2,3], **G Buza**[1], **R E Fábián**[3] and **Á Vida**[1]

[1] Bay Zoltán Nonprofit Ltd. for Applied Research, Kondorfa Street 1, Budapest, 1116, Hungary

[2] Department of Innovative Vehicles and Materials, GAMF Faculty of Engineering and Computer Science, John Von Neumann University, Izsáki Street 10., Kecskemét, 6000, Hungary

[3] Doctoral School on Materials Sciences and Technologies, Óbuda University, Népszínház Street 8., Budapest 1081, Hungary

E-mail: Adam.Vida@bayzoltan.hu

Abstract. Additive technologies are commonly used for component production and repair. Laser metal deposition is a type of additive technology that can create new bodies or add new components to existing parts. This technology enables continuous modification of the chemical and material properties of the deposited layer to suit the intended purpose. The alloying powder is dissolved and mixed in a small volume molten pool during laser deposition. The laser cladding process is known for its rapid heating and cooling rates. This can result in partial or non-existent dissolution of powder particles in the molten pool, or the formation of non-equilibrium microstructures that cannot be achieved through conventional metallurgical processes. As a result, this method can be used to create materials with unique combinations of properties. In this study, a Co-Cr coating layer was applied to a structural steel base plate. The parameters combination was generated using Taguchi's experimental design. The relationship between the parameters and the properties of the prepared coating was evaluated using the Taguchi method.

1. Introduction

Laser beam technologies are widely used in various fields of research and industry. Numerous surface modification technologies are currently in use, with new ones constantly emerging. Laser metal deposition (LMD) is a type of additive manufacturing technology. It involves melting the base material using a laser beam to create a molten pool. A different material is added to the molten pool to create a deposited layer. The filler and the base material are mixed in small quantities depending on the process parameters. This process allows for the creation of coatings with special properties, such as wear or corrosion resistance. Filler materials for laser metal deposition can be fed to the molten pool in either powder or wire form. Metal powders are used not only in laser beam additive technologies, but also in fused deposition modelling and metal injection moulding. They have various industrial applications, including maintaining and repairing worn or damaged parts, modifying surfaces to create unique properties such as wear and corrosion resistance, and producing special one-off parts, such as those used in the aerospace industry. The research work used laser metal deposition technology with metal powder. This technology applies a uniform and solid layer on the surface while minimising base material deformation and heat treatment problems. The technology achieves this through focused heat input from the laser beam. The technology is efficient in powder utilisation, and any unused powder can be reused after collection. The metal powder can be fed precisely, and different compositions can be mixed seamlessly. This allows for the creation of layers with varying properties without abrupt changes. Ongoing research is focused on developing powder and wire filler materials, investigating various powder mixtures and coatings for future advancements [1-9].

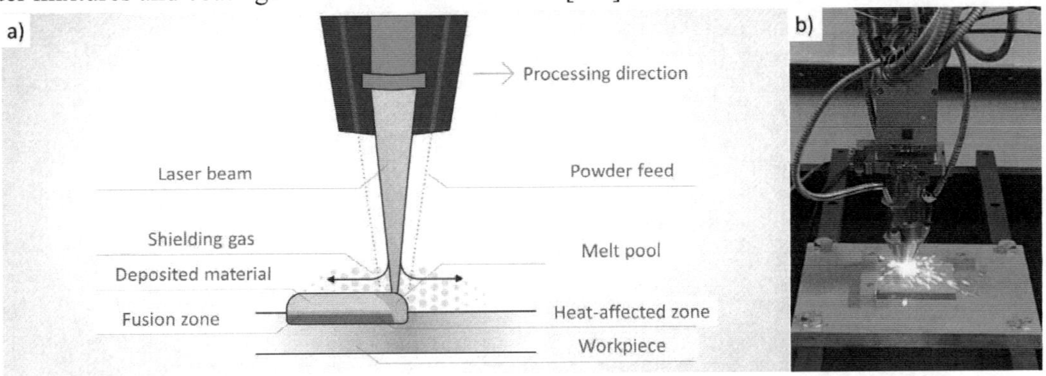

Figure 1. a) Schematic figure b) experimental setup of the Laser Metal Deposition process[10]

2. Description of the Experiment

2.1. Materials

Co6.33 alloy metal powder was used for the LMD process. The average chemical composition is presented in Table 1.

Table 1. Chemical composition of the metal powder

Material	Cr%	W%	Si%	C%	Ni%	Fe%
Co6.33	28,4	4,5	1,1	1,1	1,4	0,8

The material is a regularly shaped, gas atomized metal powder, which diameter is ranging from 50 to 150 µm. The substrate used was hot-rolled steel plates with dimensions of 30 mm x 50 mm x 12 mm. Prior to the process, the substrate surface was grinded to remove the mill scale.

2.2. Laser equipment and method

Experiments were conducted using a Trumpf 7020 cell equipped with a MSO70 LMD head and a 4 kW laser source. The LMD head was equipped SO12 nozzle. For the powder feeding we used a GTV PF/2. The nozzle has a 12 mm focal length and during the process we positioned the head so the powder feeding is on the substrate. The laser diameter is variable, and it was set to 4 mm. Argon 4.6 was used as both the shielding and carrier gas, with a gas flow rate of 8 l/min. The scanning distance was the same for all parameters. Figure 2 shows a prepared sample and deposition path.

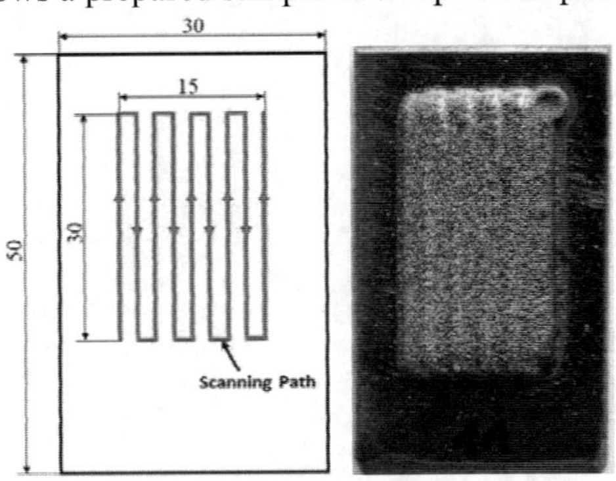

Figure 2. Single-layer sample in the top view.

The experiment was designed and analysed using Minitab 20. The experiment was designed to determine the effect of three different parameters on the investigated characteristics. We tested 5 different values per parameter, resulting in a 3-factor, 5-level design testing 25 combinations.
The evaluations varied according to the characteristics tested, depending on whether the characteristic was set to be minimised or maximised. The samples were divided into 3 separate layers for each property tested, e.g. coating layer, transition or inclusion zone, allotropic transformation. We determined the optimal laser parameters which is summarized in Table 2.

Table 2. Taguchi test input parameters

Factors	Levels
Power [W]	1500; 1750; 2000; 2250; 2500
Feed rate [m/min]	0,4; 0,55; 0,7; 0,85; 1
Powder feed rate [g/min]	11,1; 12,8; 14,5; 16,0; 17,7

3. Results and discussion

3.1. Optical microscopy results

For all 25 combinations, the applied layer was cut in half perpendicular to the direction of application, at a distance of 15 mm from the edge of the applied layer. Cross-sectional metallurgical samples were prepared, and the etched microscopy images are shown in Figure 3. Two samples were selected, 10 and 20, which best represent the difference between the parameters. The lower laser power in part a) of Figure 3 resulted in less melting of the base plate than in part b). In part a), it is possible to observe waviness, which allows for conclusions to be drawn regarding the scanning distances and degree of overlap. Different grain structures have been formed in the heat affected zone and are shown in Figure 4.

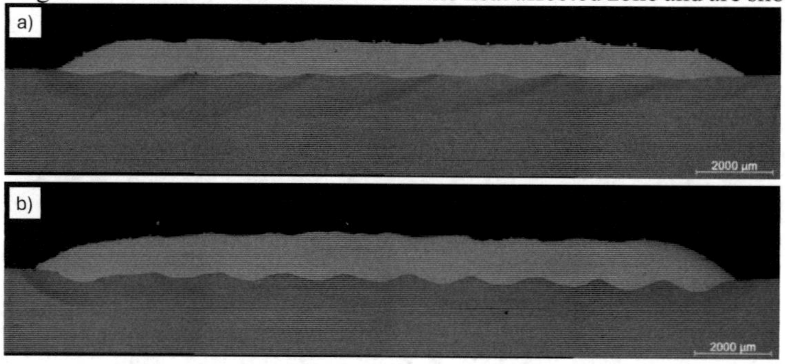

Figure 3. Schematic figure of the Laser Metal Deposition process.

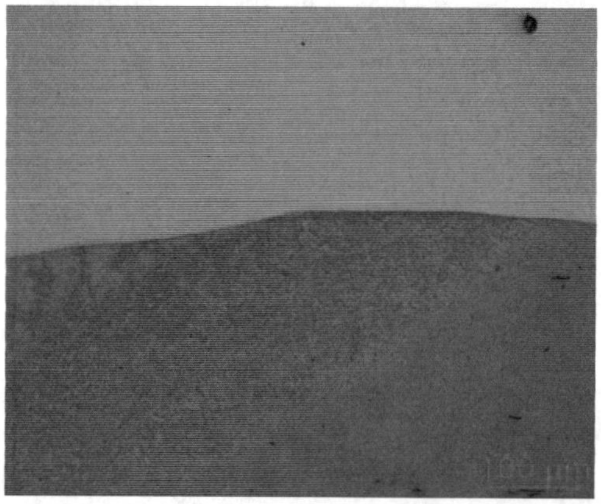

Figure 4. Structural image of the applied layer and the heat-affected zone at 200x magnification

The samples were etched with 2% NITAL solution, as it was important to see how the boundary layer was in contact with the coating. No defects, porosities or inclusions were found in the boundary layer. The coated layer's microstructure is dendritic, with minimal porosity, and the largest porosity defect measures 20 µm, as shown in Figure 4. The well-defined boundaries of the heat affected zone are shown in Figure 3, an enlarged image of which is shown in Figure 4. The change in microstructure is also visible, consisting fine-grained and soft martensite.

3.2. Taguchi results

To determine the optimal parameter for the deposited layer thickness, the "larger is better" evaluation method was ideal for us. A thicker deposited layer results in a greater wear layer and increased material thickness for subsequent machining processes. The resulting diagram is shown in Figure 5.

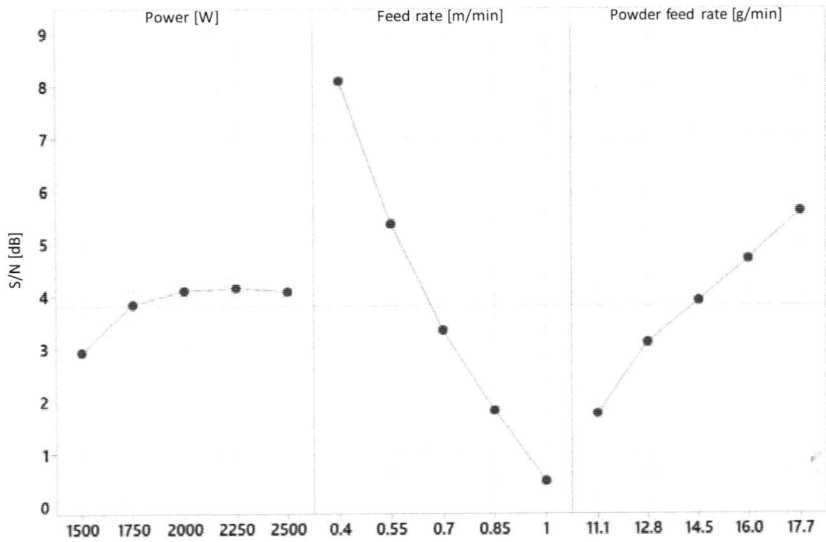

Figure 5. Signal/noise diagram for applied layer thickness.

The diagram shows that the power parameter curve reaches its maximum at 2250W, but further increases have a diminishing effect on the thickness of the deposited layer. A slower feed rate results in a thicker applied layer, as expected. Moreover, a more intensive powder feed leads to a thicker layer. In summary, the laser feed rate and the rate of the powder feeder greatly affect the signal-to-noise ratio.
To achieve the maximum thickness of the deposited layer, it is recommended to use a parameter triplet of 2250 W, 0.4 m/min, and 17.7 g/min. Excessive power should be avoided as it can cause the layer to re-melt excessively, which is unfavourable for achieving maximum thickness.
The hardness measurement results of the applied layer were evaluated using the 'bigger is better' method. A search was made for parameters that would result in a harder coating. Figure 6 shows the signal/noise diagram resulting from the evaluation.

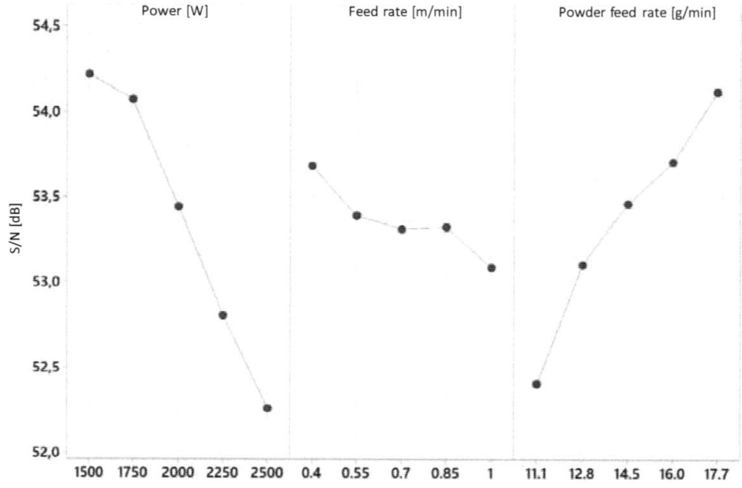

Figure 6. Signal/noise diagram for the hardness of the applied layer.

The diagram shows that the harder layers are produced at lower power and higher powder feed rates and are less affected by the laser head feed rate. The lower power results in less heat input and smaller molten pool which results a faster cooling rate. Higher powder feed rates cool the molten pool more intensively, which is reflected in the hardness values. The feed rate has less influence on the layer hardness. At the lowest power and the highest powder feed rate, the hardness of the layer is 510-520 HV. The points tested are closely approximated by a straight line, no major differences are observed.

4. Conclusion

The LMD experiments successfully produced a uniform solid coating layer on a hot-rolled steel substrate using Co6.33 metal powder. The thickness of the applied layer was found to be consistent due to the good overlap between the layers. The effect of the process parameters was investigated by Taguchi experimental design. It was possible to determine from the experiments which process parameters affect the thickness of the coating layer. From this it was possible to determine laser feed rate and powder feed rate. The power was not an affecting factor. However, the microscopic image shows that increasing the power increases the mixing of the substrate to the layer. The feed rate has the least effect on the hardness of the coating layer. Increasing the power leads to a decrease in the hardness of the coating layer. Increasing the powder feed rate increases the hardness.

References

[1] Berczeli M., Tajti F., Juhász G., Weltsch Z. 2024 Opt Laser Technol, **174** 110556
[2] Kovács P, Körömi B, Weltsch Z and Berczeli M 2023 *Mater.* **16**, 7663
[3] Berczeli M and Weltsch Z 2018 *IOP Conf Ser Mater Sci Eng.* **448** 012050
[4] Berczeli M. 2020 *Acta. Phys. Pol. A.* **137** 864-7
[5] Berczeli M and Weltsch Z 2021 *Polymers.* **13(6)** 901
[6] Bian L., Shamsaei N., Usher J. 2022 Laser-Based additive manufacturing of metal parts: Modeling, Optimization, and Control of Mechanical Properties *CRC Press* pp 2-23
[7] Ocelík V., Hosson,D. J. 2018 Advances in laser materials processing. *(Elsevier eBooks)* Chapter 17-19 pp 507-94
[8] Ledniczky G., Weltsch Z. 2023 *Acta Mater. Trans.*, ***6*/1**, 33–7
[9] Cheng J., Xing Y., Dong E., Zhao L., Liu H., Chang T., Wei J. 2022 *Materials*, ***15*(16)**, 5522
[10] Marco G. 2019 Overview of Laser Metal Deposition Applications, Trends and Products *Int. Conf. on Addit Manuf.*

IOP Conf. Series: Materials Science and Engineering **1313** (2024) 012017 doi:10.1088/1757-899X/1313/1/012017

Characterization of the hydrostatic core and plastic zone that form during Brinell and Vickers hardness tests

F. Oláh [12], M. Réger [1], R Horváth [1*]

[1]Óbuda University, Donát Bánki Faculty of Mechanical and Safety Engineering, H1081, Népszínház u. 8., Hungary
[2]Doctoral School on Materials Sciences and Technologies, Óbuda University, Budapest
*horvath.richard@uni-obuda.hu

Abstract. During a hardness test, a hydrostatic core and a plastic zone are formed under the impression due to the load. The size and shape of the hydrostatic core and the plastic zone are greatly influenced by the tested material, the loading force and the shape of the indenter. In this study, we examine in detail the characteristics of the hydrostatic core and plastic zone formed during a Brinell and Vickers hardness test with the same (bilinear) hardening material model and with the same load using the finite element method in the ANSYS environment. The results of our resarch is concluded that the characteristics under the indenter can be determined well with finite element modeling and in the case of the Vickers procedure, the ratio of the hydrostatic core to the plastic zone is greater for all loads as in the case Brinell method.

1. Introduction

The conditions under the indenter during a hardness test can be described with the mechanical model of the expansion of the cavity, which is assumed to be spherical. Bishop et al.[1] were among the first to examine the theory of indentation during hardness testing. Johnson[2] examined and described the formation of the of the hydrostatic core and plastic zone (Figure 1) under the indenter. In his expanding cavity model, he assumed that under the indenter during loading, there is a hemispherical zone of hydrostatic stress state, the so-called "hydrostatic core". And around this, a plastic deformation zone is created around the hydrostatic core, where stresses and displacements can be characterized by radial symmetry. During the hardness test, the size of the hydrostatic and plastic zone increases as the load increases. In this approach, the hydrostatic core basically plays the role of load transfer. The load is absorbed by the plastic zone, which deforms under pressure. Figure 1 illustrates the conditions under a spherical penetrating body (Brinell hardness test). Knowledge of the plastic zone is very important, as it is necessary, among other things, in determining yield strength and, also, for example, the influencing effect of thin coating layers during indentation [5]. Johnson's model [2] therefore provides a relationship between hardness, the hydrostatic core, the size of the plastic zone, and the elastic-plastic mechanical properties of the material. With Johnson's and later Lawn et al.'s study, it became possible to predict the size of the plastic zone (Figure 1) [2, 6, 11]. The size of the hydrostatic core and plastic zone created during indentation also greatly depend on the geometric characteristics of the indenter. In this article, we examine Brinell (spherical indenter) and Vickers (pyramidal indenter) impressions in detail, using the same conditions, with the finite element method, which was later validated with actual hardness tests. During the comparison, we examine the size and shape of the hydrostatic cores and the plastic zone.

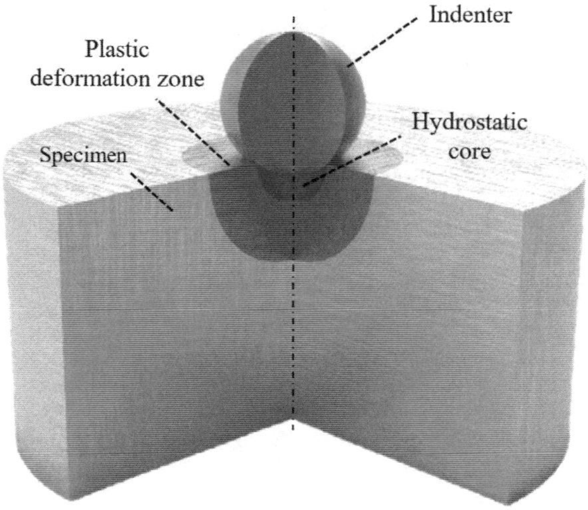

Figure 1. Representation of the hydrostatic core forming under a spherical (Brinell) indenter and the plastic and elastic zones (based on [2] [3][4])

2. Materials and methods

The tests were performed on S235JR (1.0045) structural steel with loads of 30 and 187.5 kg for Brinell [7] and Vickers [8] indenters. Finite element modeling in the Ansys environment can be validated from the obtained prints and indentation depth. A so-called bilinear (hardening) material model was used for finite element modeling (Figure 2.) – hardening modulus is the slope of the line drawn between yield strength and tensile strength [9, 13] –, whose values were determined from the averages of three tensile tests. The yield strength of the tested material $\boldsymbol{\sigma}_{ys}$ = 332±2.3 MPa, and its tensile strength is $\boldsymbol{\sigma}_{m}$ = 590.4±6.9 MPa. The elasticity and hardening (tangent) modulus were calculated from the associated values of stress and deformation as follows:

$$E = \frac{\sigma_{ys}}{\varepsilon_{ys}} \tag{1}$$

$$E_T = \frac{\sigma_m - \sigma_{ys}}{\varepsilon_m - \varepsilon_{ys}} \tag{2}$$

where E = modulus of elasticity, MPa; σ_{ys} = yield strength, MPa; ε_{ys} = elastic deformation up to the moment of yield, mm; E_T = tangent modulus, MPa; σ_m = tensile strength, MPa; ε_m = deformation at tensile strength, mm.

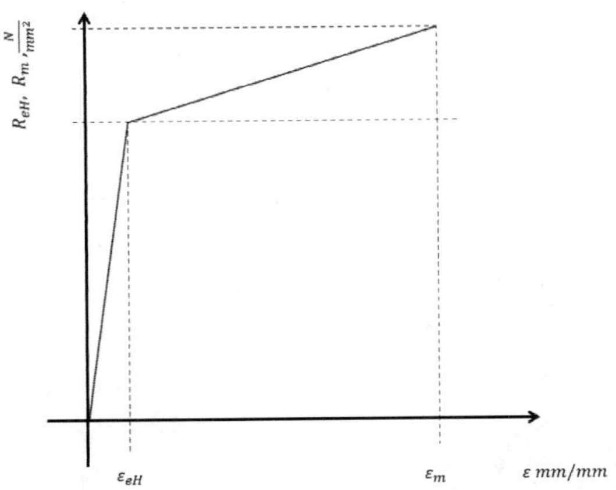

Figure 2. The bilinear hardening model used in the finite element environment

Due to the geometric characteristics of Vickers hardness testing, it was necessary to create 3D bodies in the finite element ANSYS environment. In other words, 2D mapping was not possible, and accordingly, we modeled spatial quarters in both cases. This way, we were able to model Brinell and Vickers indentation in a comparable way. During meshing, in the case of Vickers, a 0.05 mm tetrahedral mesh was created within 2 mm of the apex of the pyramid. Although a tetrahedral mesh requires more computing capacity than a hexahedral mesh, it calculates the stresses at several points. In the case of the Brinell indenter, a 0.01 mm hexahedral meshing was set in the vicinity of 1 mm of contact, which thus increased the fineness of meshing. The different meshing is necessary because of the different deformations, as can be seen there is a difference in the result according to the depth and volumes, and in the case of the Vickers pyramid, more slip planes are needed because of the pointed body. The meshes of the tools and workpieces are identical and are concentrated around the tool head as described. The tools have the material properties of diamond, so no rigid bodies were used in the process. The coefficient of friction was set to 0.12, as previous studies have shown that there is no real difference in the computational cases of different frictions.

The load during indentation primarily causes permanent deformation in the plastic zone. Accordingly, in the loaded state, the hydrostatic core (3) and plastic zone (4) can be determined based on the stress and deformation states at the nodes of the finite element model as follows:

- The hydrostatic core is the volume where during loading, the first scalar invariant of the stress tensor (I_1) at the nodes of the finite element model is greater than the yield strength of the material, i.e. the following condition is met:

$$I_1 = \frac{\sigma_1 + \sigma_2 + \sigma_3}{3} \geq \sigma_{ys} \tag{3}$$

where $\sigma_1, \sigma_2, \sigma_3$ – the values of the main diagonals of the stress tensor in the finite element nodes, MPa.

- The plastic deformation zone can be defined from the set of nodes that have a plastic deformation at the maximum load, i.e. the following equation is fulfilled [10]:

$$\varepsilon_0^P = \frac{\sqrt{2}}{3}\left[\left(\varepsilon_x^P - \varepsilon_y^P\right)^2 + \left(\varepsilon_y^P - \varepsilon_z^P\right)^2 + \left(\varepsilon_z^P - \varepsilon_x^P\right)^2 + 6\left(\varepsilon_{xy}^P\right)^2 + 6\left(\varepsilon_{yz}^P\right)^2 + 6\left(\varepsilon_{xz}^P\right)^2\right]^{\frac{1}{2}} \geq 0 \tag{4}$$

where, ε_0^P = equivalent plastic strain, mm/mm; $\varepsilon_x^P, \varepsilon_y^P, \varepsilon_z^P, \varepsilon_{xy}^P, \varepsilon_{yz}^P, \varepsilon_{xz}^P$ = components of the plastic strain tensor, mm/mm.

During the tests, it is important to determine the volume of the zones formed in this way (V_h – hydrostatic core, mm^3 and V_p – plastic zone, mm^3).

3. Characteristics of Brinell and Vickers hardness tests

Figure 3 shows the top view of the 30 and 187.5 kg impressions. For Brinell testing, the averages of the diameters of the impressions are 0.53 and 1.23 mm. In the case of the Vickers test, the averages of the diagonals were 0.624 and 1.535 mm.

It was also possible to examine the impressions in 3D with the microscope images, with which the depth of the impression can also be determined. Considering the geometric calculations, the depths set in the software were 0.03 and 0.162 mm for the Brinell and 0.084 and 0.21 for the Vickers tests.

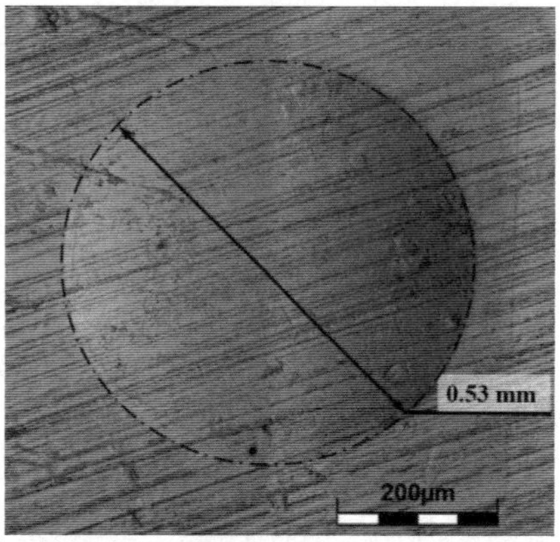

Brinell indentation impression – load 30 kg

Vickers indentation impression – load 30 kg

Brinell indentation impression – load 187.5 kg

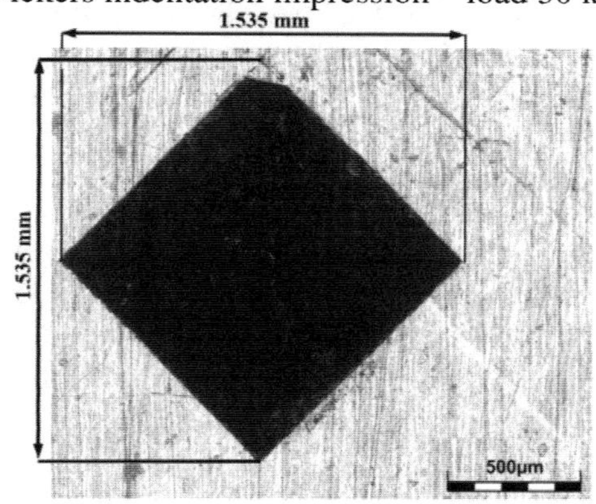

Vickers indentation impression – load 187.5 kg

Figure 3. Top view of Brinell and Vickers impressions

4. Analysis of the hydrostatic core and plastic zone in an Ansys environment

The volume of the zones formed during the finite element simulations is summarized in Table 1, which also contains the ratio of the zones.

Table 1. Volumes of zones formed under Brinell and Vickers impressions.

	Load 30 kg		**Load 187.5 kg**	
	HB	**HV**	**HB**	**HV**
V_h, mm³	0.165	0.105	3.06	1.916
V_p, mm³	0.729	1.086	17.26	14.922
V_h/V_p	4.4	10.3	5.64	7.78

The difference in the zones can stem from several reasons. The recorded data show that, in the case of Vickers hardness, a larger displacement was required to absorb the same load, and the shape deviations are clearly visible from the deformation volume obtained. In the Vickers test, it is clear that the stress distribution along the edges and faces is different, therefore the part suffering permanent deformation has larger protruding parts, while in the case of the Brinell, the circular symmetrical distribution results

IOP Conf. Series: Materials Science and Engineering **1313** (2024) 012017 doi:10.1088/1757-899X/1313/1/012017

in a more even load transfer, which is why a smaller absorbing part is probably enough for it. Thus, based on finite element runs, we selected finite element nodes that satisfy equation (3) or (4). Based on the coordinates of the selected nodes (point clouds), the volumes were defined and displayed in another modelling software. These zones are presented in Figure 4.

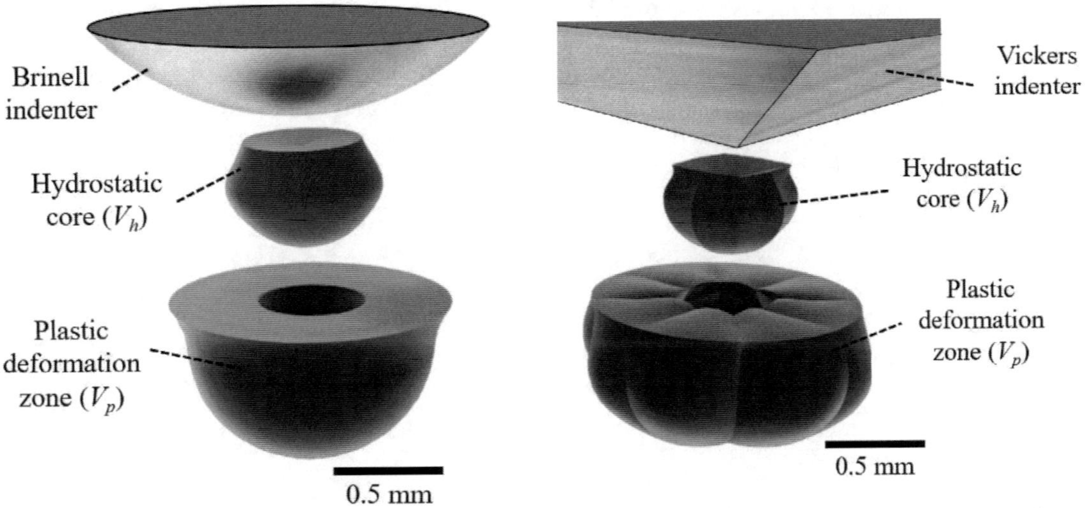

Figure 4. Zones as a result of a load of 30 kg

5. Examining the distance between indentation impressions

The distance between successive hardness tests is prescribed by a standard (ISO 6506-1 – for the Brinell procedure, ISO6507 – for the Vickers procedure) [7, 8, 12]. The standard prescribes 2.5 times the size of the expected impression from the edge of the specimen, and usually 3 times the diagonal or diameter of the impression between the impressions. Based on the results of the finite element tests, this recommendation can be verified from the size of the impressions and the deformation zone around them. Based on the top view images of the impressions and the maximum contour of the deformation zone (Figure 5), their ratios are ~2.5 in both cases (Brinell impression diameter: 0.53 mm, diameter of the deformation zone: 1.32 mm; Vickers diagonals: 0.624 mm, and the diameter of the circle around the deformation zone: 1.544 mm). The results obtained from the finite element simulations also confirm the recommendation of the standard concerning the distances between the impressions.

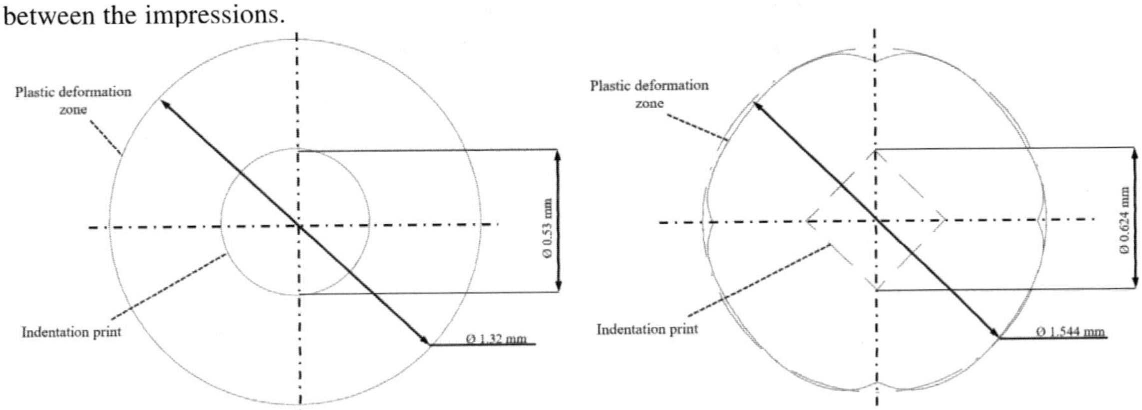

Brinell impression and deformation zone for a load of 30 kg

Vickers impression and deformation zone for a load of 30 kg

Figure 5. The maximum size of the deformation zone under impressions

6. Conclusion

In this article, we performed a finite element simulation of two Brinell and Vickers impressions created with the same load on the same material, to determine what size and shape hydrostatic and plastic zones are formed under indenters with different geometries. The zones were characterized by their volume and ratio. During the finite element tests, we used a so-called bilinear hardening material model, which was determined from the actual yield strength and tensile strength of the tested steel. Based on the simulations, we can conclude that the characteristics under the indenter can be determined well with finite element modeling, and in the case of the Vickers procedure, the ratio of the hydrostatic core to the plastic zone is greater for all loads, due to the geometric properties of the indenter. Based on the dimensions of the plastic zones, finite element modeling also confirmed the recommended distances between impressions.

Acknowledgments

Supported by the ÚNKP-23-3 New National Excellence Program of the Ministry for Culture and Innovation from the source of the national research, Development and Innovation Fund.
This research was funded by the 2020-1.1.2-PIACI-KFI-2020-00129 project. The authors acknowledge the financial support of this work by Hungarian-Japanese bilateral project (2019-2.1.11-TÉT-2020-00204.)

7. References

[1] Bishop, R. F., Hill, R., & Mott, N. F. (1945). The theory of indentation and hardness tests. *Proceedings of the Physical Society*, *57*(3), 147. https://doi.org/10.1088/0959-5309/57/3/301
[2] Johnson, K. L. (1970). The correlation of indentation experiments. *Journal of the Mechanics and Physics of Solids*, *18*(2), 115-126. https://doi.org/10.1016/0022-5096(70)90029-3
[3] Marsh, D. (1964). Plastic flow in glass. Proceedings of the Royal Society of London. Series A. Mathematical and Physical Sciences, 279(1378), 420-435. https://doi.org/10.1098/rspa.1964.0114
[4] Tabor, D. (1970). The hardness of solids. Review of physics in technology, 1(3), 145. https://doi.org/10.1088/0034-6683/1/3/I01
[5] Mata, M., Casals, O., & Alcalá, J. (2006). The plastic zone size in indentation experiments: The analogy with the expansion of a spherical cavity. International Journal of Solids and Structures, 43(20), 5994-6013. https://doi.org/10.1016/j.ijsolstr.2005.07.002
[6] Lawn, B. R., Evans, A. G., & Marshall, D. B. (1980). Elastic/plastic indentation damage in ceramics: the median/radial crack system. Journal of the American Ceramic Society, 63(9-10), 574-581. https://doi.org/10.1111/j.1151-2916.1980.tb10768.x
[7] ISO 6506-1:Metallic materials – Brinell hardness test Part 1: Test method
[8] ISO 6507-1: Metallic Materials – Vickers Hardness Test – Part 1: Test method
[9] Amlashi, H. K., & Moan, T. (2008). Ultimate strength analysis of a bulk carrier hull girder under alternate hold loading condition–A case study: Part 1: Nonlinear finite element modelling and ultimate hull girder capacity. Marine structures, 21(4), 327-352. https://doi.org/10.1016/j.marstruc.2007.12.006
[10] Mendelson, A. (1983). Plasticity: theory and application. Krieger Publishing Company.
[11] Ma, Y., Zhang, Z., & Li, J. (2023). Finite Element Analysis of Indentation Tests for Different Hardness Measurements: A Comparative Study. Journal of Materials Science & Technology, 152, 1123-1131.
[12] Gao, H., Wang, X., & Zhou, T. (2022). Characterization and Simulation of Indentation Behavior in Metals Using Advanced Finite Element Modeling. Materials & Design, 215, 110515.
[13] Kim, S., & Lee, D. (2024). Comparative Study on Hydrostatic and Plastic Zones in Hardness Testing: Experimental and Numerical Approaches. International Journal of Mechanical Sciences, 239, 107500.

IOP Conf. Series: Materials Science and Engineering **1313** (2024) 012018 doi:10.1088/1757-899X/1313/1/012018

Improving the bonding technology of aluminium sheet used in the construction industry by gas flame surface treatment

F Tajti[1] and A Capár[1]

[1] Department of Innovative Vehicles and Materials, GAMF Faculty of Engineering and Computer Science, John Von Neumann University, Izsáki Street 10, Kecskemét, Hungary

E-mail: tajti.ferenc@nje.hu

Abstract. According to current building trends, aluminium sheets can be used as building cladding elements. The advantage is that aluminium has a low density but high strength and resistance to atmospheric corrosion. For a nice appearance, these sheets are coloured with a polyester-based powder coating and are fixed by adhesive bonding on the reverse side. For adhesion, a flexible adhesive is required, and it should be noted that the surface is painted. From a safety point of view, it is important that the strength of the bonded joint does not deteriorate over many years. In our research, we investigate the applicability of gas flame surface treatment technologies to improve the bonding technology of aluminium sheets coated with polyester-based paints. The effectiveness of the surface treatment is tested by surface wettability measurements and, based on the results obtained, test joints are made to test their tensile and shear strength.

1. Introduction

Developments in the construction industry have resulted in new types of building cladding [1]. New types of materials must meet many requirements. Cladding elements are constantly exposed to environmental influences such as sun, heat, cold, rain, wind and salt [2]. In some cases, they may come into contact with chemicals such as cleaning products. Some metals and metal foams can withstand these stresses very well and not to forget, they also provide an aesthetically pleasing appearance. The most used metal for this purpose is aluminium [3]. For an even better appearance or for corrosion protection, these metal elements can be coloured. The best solution for this area and use is to apply polyester-based powder paints, as they adhere well to the surface of the metal, protecting and decorating it for many years [4,5]

The attachment of the elements to the building facade is often not done with the traditional fastening techniques such as screwing [6] and riveting [7], soldering [8], laser welding [9] because they degrade the appearance [10]. To improve the appearance, aluminium profiles are adhered to the back of the cladding elements and screwed to a bracket [11,12]. Because of the stresses listed above, the bonding of metal surfaces must be well planned [13,14]. Special flexible adhesives are needed to ensure long-lasting fixation, adhesion of the painted surface and environmental influences. High-strength bonding of pure metals in complex applications requires a lot of precise preparation, even when using special adhesives [15-18]. Even in such cases, surface preparation is justified. There are several methods for this, such as surface roughening by grinding or sanding, chemical surface preparation, but also high energy density technologies such as laser or plasma surface treatment [19-21]. However, bonding painted surfaces and polymers requires a different kind of preparation [22]. In such cases, researchers have achieved good results in increasing surface adhesion using PB gas flame treatment [23].

2. Materials and Methods

For the surface treatments, 3105 aluminium alloy was used and coloured with a 30-50 μm thick polyester based paint which we prepared in 25 x 75 mm^2 pieces. This material is often used as a building cladding element, which is why it is a priority to improve the adhesive technology for this type of painted surface.

Sikaflex-252 adhesive was used to ensure a flexible bond. This adhesive normally requires extensive surface preparation consisting of a surface sanding, alcohol cleaning, primer application and finally application of the adhesive of a specified thickness. For this reason, we want to implement a surface treatment process that avoids these lengthy preparation operations.

The gas flame treatment was carried out with propane butane gas burned with a bunsen burner. In order to repeat the surface treatment, the sheet to be treated was moved above the gas flame at a defined height and speed by an ABB robot, as shown in Figure 1. Variable parameters between treatments were the distance from the top of the bunsen burner to the Al sheet and the speed of movement of the Al sheet over the flame. During the surface treatment of the plates, the aim was to ensure that the gas flame would not cause any visible damage to the surface.

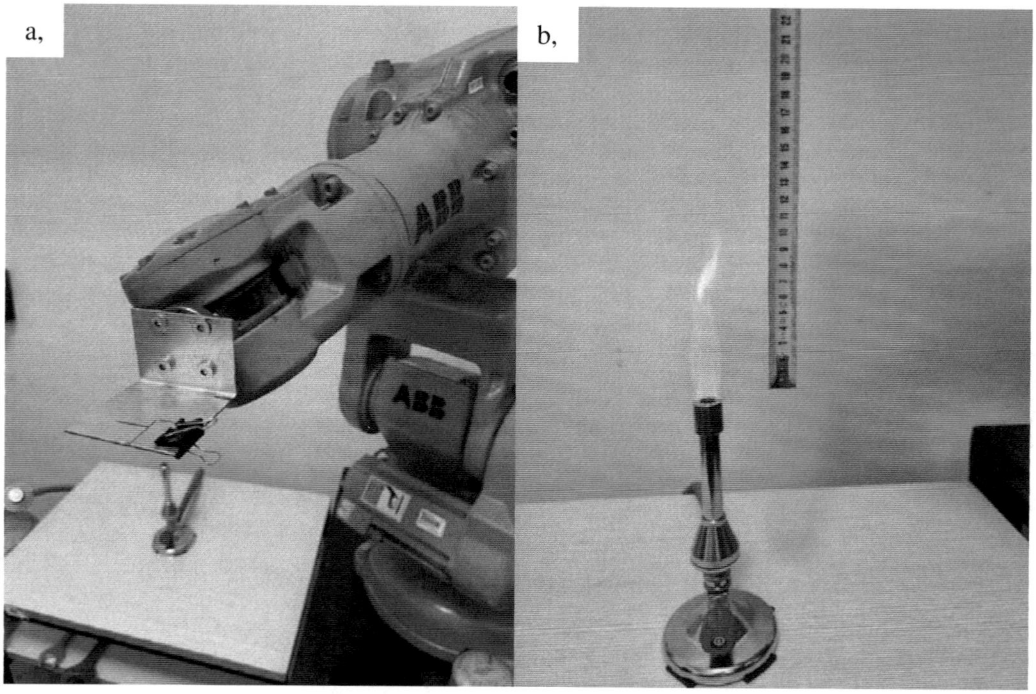

Figure 1. a) ABB robot with sheet gripper, b) bunsen burner

Gas flame treated surfaces are subjected to a wetting test. The sample is placed on the vertically movable table of the wetting tester. A micropipette is used to drop liquid onto the sample surface. The liquids should be applied to the surface with as little kinetic energy as possible, so the liquid is not dropped on the surface, but is squeezed out onto the tip of the pipette and the sample is lifted towards it. In the test, 5 μl of distilled water and ethylene glycol are added to the surface. A side view photograph of the droplet was taken immediately after the test was completed.

The contact angles are measured on the finished photographs. The surface of the workpiece is determined in the measuring software and a tangent is drawn from the two contact points of the liquid drop to the drop. The angle between the sample surface and the tangent drawn to the drop is called the wetting contact angle, as shown in Figure 2. The contact angles of the two sides may be asymmetric, so the average of the two contact angles was taken to evaluate the measurements.

Figure 2. Contact angle measurement

Using the contact angle values obtained with the two liquids and the Fowkes equation, we can calculate the surface energy of the sheet after various surface treatments.

After evaluation of the wetting measurement, where we observed a high degree of improvement in adhesion and considered the treatment parameters to be industrially usable, we prepared bonded samples for tensile and shear stress tests.

Three types of sample bonding have been completed. One control sample, cleaned only with alcohol, one according to the Sika technology recommendation and finally the bonding of the alcohol-cleaned and gas flame treated sheets. All bonding tests were performed according to ASTM D1002. The adhesive thickness was adjusted to 3 mm using double-sided spacer tape and the bonded surface was 12,5 x 25 mm^2.

3. Results

The surface energy of the untreated sheet was calculated to be 20 mN/m based on the wetting measurement. This low value means poor adhesion between the surface and the adhesive. It can be seen that this must be increased in order to create a high strength and durable bond on the surface. Figure 3 shows the results obtained with gas flame treatment on the aluminium sheets.

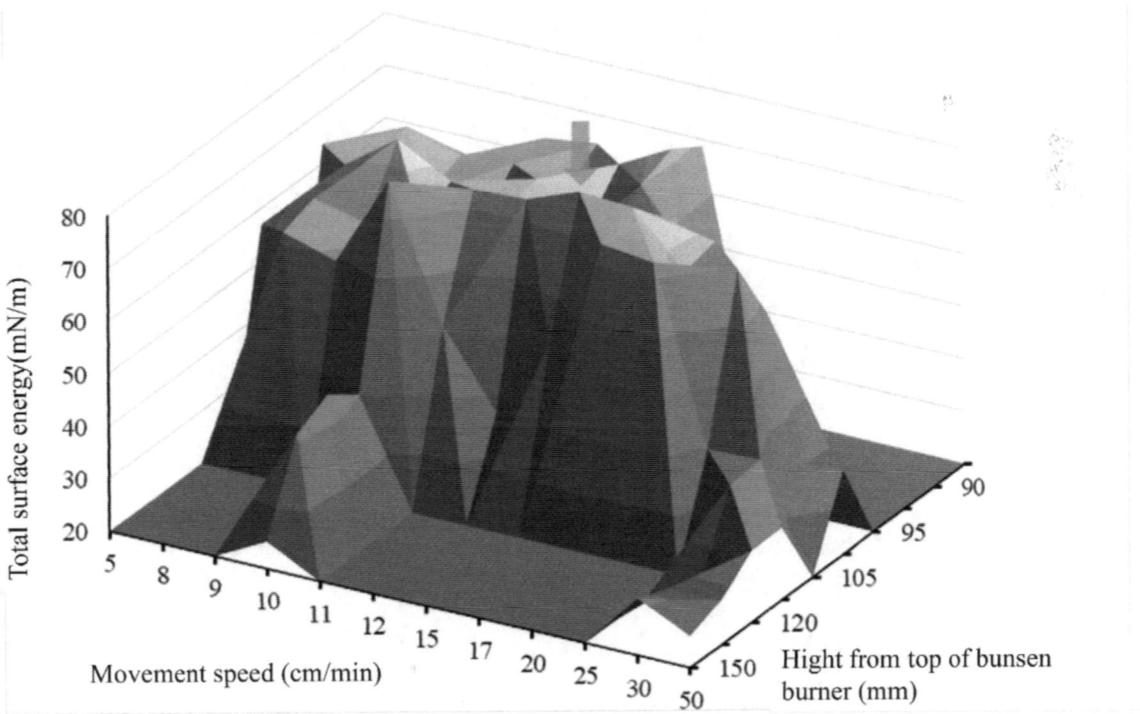

Figure 3. Effect of gas flame treatment on the surface energy of aluminium

It can be seen that we found several settings where the surface energy increases significantly, but we had to choose one that is also used in industry, so it is fast and far enough away from the surface to be treated. The best surface treatment was found to be 200 mm from the bunzen burner and a movement speed of 15 cm/min. At this point the surface energy is above 80 mN/m which is already optimal for bonding technologies.

For the bonding tests we prepared the specimens in two ways, for the shear tests we made overlapped joints and for the tensile tests we glued an L-section to the sheet and performed the tensile tests on these. The tensile tests were carried out on three different parameters: the control sample (SA), the case bonded with Sika technology (SK) and the gas flame treated case (SL). There was a case where adhesion failure was observed, as shown in Figure 4, in this case the adhesive was peeled off the surface in one layer.

Figure 4. Adhesive failure

While Figure 5 shows a cohesive failure, where the adhesive adheres to the surface so well that the adhesive splits in two during the failure. In this case, we can use the full strength of the adhesive.

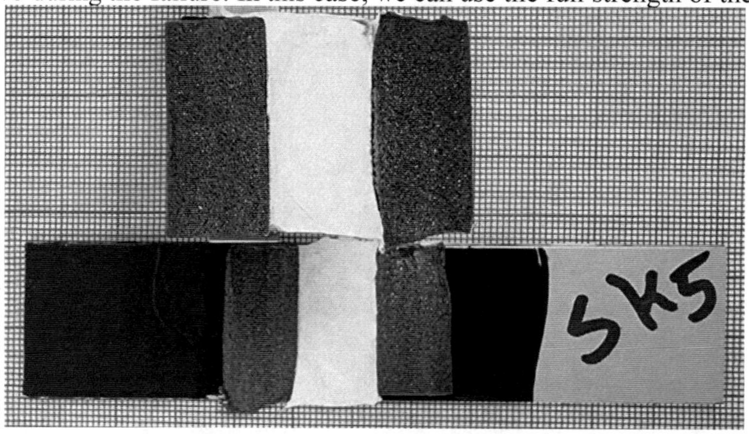

Figure 5. Cohesive failure

Figure 6 shows the average of the results of the tensile tests. It can be seen that there is no significant difference in tensile strength between the three types of samples. When bonding a large plate, strength would not be a problem. However, in the case of the sheet cleaned with alcohol only, adhesion failure has occurred, so its value may change.

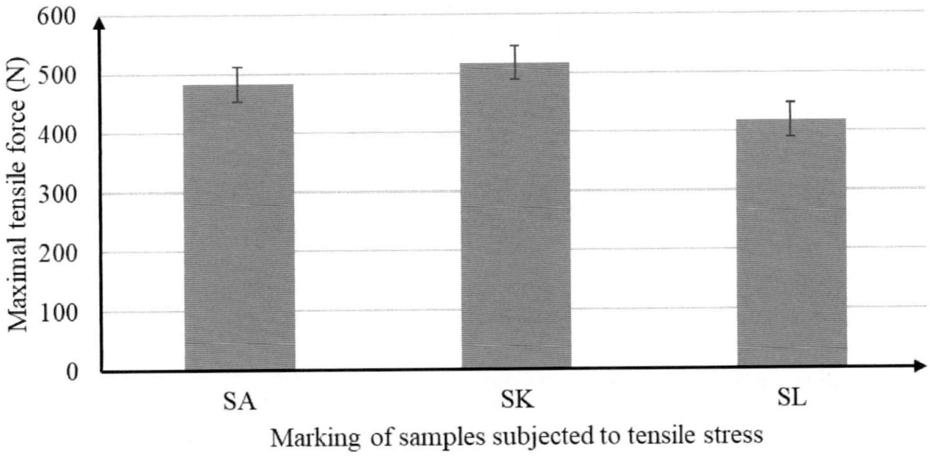

Figure 6. Results of tensile tests

The results of the shear tests are shown in Figure 7, and there is not much difference in the maximum of shear force between the different samples. However, again it was observed that the plates cleaned with alcohol only also showed adhesion failure, while the other two showed cohesive failure.

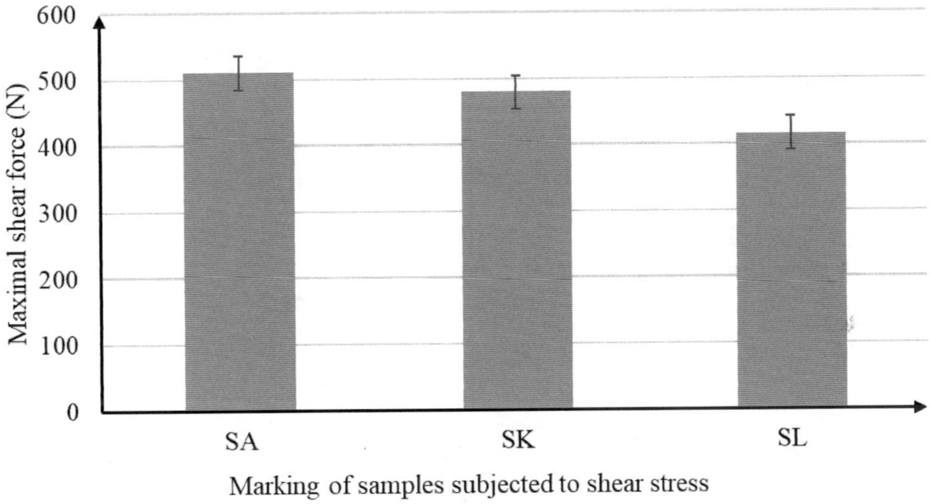

Figure 7. Results of shear tests

4. Conclusion

Aluminium sheets painted with a polyethylene-based paint require surface treatment. A good solution is the gas flame surface treatment. The wetting results show that compared to the untreated case, where the surface energy was 20 mN/m, it was increased to over 80 mN/m. As a result, the poor adhesion failure for both shear and tensile stresses was improved to cohesive failure. However, there is no significant difference in the tensile and shear forces. Although the preparation operations recommended by the adhesive manufacturer are effective, they consist of several steps and take a long time. This is not permissible in industrial conditions and therefore gas flame surface treatment is considered preferable to this technology, which can be carried out quickly and to good quality with the chosen parameter.

Acknowledgments
This work was supported by the 2020-1.1.2-PIACI-KFI-2021-00260 Grant. The authors are grateful to the scholars cited in this paper, and their research results have given a lot of inspiration to this work. Special thanks to the staff of tour department for their support.

References

[1] Papadimitriou I, Efthymiadis P, Kotadia H R, Sohn I R and Sridhar S. Joining TWIP to TWIP and TWIP to aluminium: A comparative study between joining processes, joint properties and mechanical performance. *J Manuf Process*. 2017;30:195-207.

[2] Kumar D, Phanden R K and Thakur L. A review on environment friendly and lightweight Magnesium-Based metal matrix composites and alloys. *Mater Today Proc*. 2021;38:359-364.

[3] Berczeli M andWeltsch Z. Improvement of adhesive joining of hybrid aluminum – GFRP using surface modifications. *IOP Conf Ser Mater Sci Eng*. 2018;448.

[4] Berczeli M. Wettability Changing of FINEMET Substrates Using High-Energy Femtosecond Laser Impulses. *Acta Phys Pol A*. 2020;137(5):864-867 PG - 4.

[5] Xu P, Coyle T W, Pershin L and Mostaghimi J. Superhydrophobic ceramic coating: Fabrication by solution precursor plasma spray and investigation of wetting behavior. *J Colloid Interface Sci*. 2018;523:35-44.

[6] Skovron J D, Rohan P R, Ulutan D, et al. Effect of Thermal Assistance on the Joint Quality of Al6063-T5A During Flow Drill Screwdriving. *Journal of Manufacturing Science and Engineering, Transactions of the ASME*. 2015;137(5).

[7] Kim K Y, Sim J, Jannat N E, Ahmed F and Ameri S. Challenges in riveting quality prediction: a literature survey. *Procedia Manuf*. 2019;38:1143-1150.

[8] Hlinka J, Berczeli M, Buza G and Weltsch Z. Wetting properties of Nd: YAG laser treated copper by SAC solders. *SOLDERING & SURFACE MOUNT TECHNOLOGY*. 2017;29:69-74.

[9] Markovits T and Berczeli M. EXAMINATION OF METAL SIDE HEATING OF LAMP JOINT BY CO2 LASER BEAM. *MATERIALOVÉ INZINIERSTVO-MATERIALS ENGINEERING*. 2018;24:82-87.

[10] Berczeli M and Buza G. Relationship between the keyhole laser welding and the plasma. *IOP Conf Ser Mater Sci Eng*. 2018;448.

[11] Tímári A, Berczeli M, Líska J and Weltsch Z. Bending and tensile test of glued aluminium foams. *IOP Conf Ser Mater Sci Eng*. 2018;448:12067.

[12] Cavezza F, Boehm M, Terryn H and Hauffman T. A Review on Adhesively Bonded Aluminium Joints in the Automotive Industry. *Metals 2020, Vol 10, Page 730*. 2020;10(6):730.

[13] Berczeli M and Weltsch Z. Mechanical properties of different strength steels MIG brazed joints. In: *34th International Colloquium on Advanced Manufacturing and Repairing Technologies in Vehicle Industry*.; 2017:29-32.

[14] Kovács P, Körömi B, Weltsch Z and Berczeli M. Investigation of the Tensile Strength of Adhesive-Bonded Steels Using Surface Treatments. *Materials 2023, Vol 16, Page 7663*. 2023;16(24):7663.

[15] Berczeli M and Weltsch Z. Enhanced wetting and adhesive properties by atmospheric pressure plasma surface treatment methods and investigation processes on the influencing parameters on HIPS polymer. *Polymers (Basel)*. 2021;13(6):901.

[16] Wu Y, Lin J, Carlson B E, et al. Effect of laser ablation surface treatment on performance of adhesive-bonded aluminum alloys. *Surf Coat Technol*. 2016;304:340-347.

[17] Cui C and Liu W. Recent advances in wet adhesives: Adhesion mechanism, design principle and applications. *Prog Polym Sci*. 2021;116:101388.

[18] Da Rosa WLDO, Piva E, Da Silva A F. Bond strength of universal adhesives: A systematic review and meta-analysis. *J Dent*. 2015;43(7):765-776.

[19] Liu M, Li M T, Xu S, Yang H and Sun H B. Bioinspired Superhydrophobic Surfaces via Laser-Structuring. *Front Chem*. 2020;8(October):1-6.

[20] Heitz J, Reisinger B, Fahrner M, Romanin C, Siegel J and Svorcik V. Laser-induced periodic surface structures (LIPSS) on polymer surfaces. *International Conference on Transparent Optical Networks*. Published online 2012:1-4.

[21] Berczeli M. Effects of CO2 laser equipment technology parameters on different materials during cutting and engraving processes. *PERNER'S CONTACTS*. 2019;2(19):7-41.

[22] Berczeli M, Kürtösi R, Tímári A and Líska J. Optimize the joining technology of CFRP and aluminium sheets using hybrid and high strength adhesive. *IOP Conf Ser Mater Sci Eng*. 2018;448.

[23] Lu J, Huang T, Liu Z, Zhang X and Xiao R. Long-term wettability of titanium surfaces by combined femtosecond laser micro/nano structuring and chemical treatments. *Appl Surf Sci*. 2018;459(January):257-262.

IOP Conf. Series: Materials Science and Engineering **1313** (2024) 012019 doi:10.1088/1757-899X/1313/1/012019

Adhesive Bonding of Carbon Fiber Reinforced Polymer Composite

P I Kovács[1,2]

1. John von Neumann University, Kecskemét, Hungary
2. Széchényi István University, Győr, Hungary

E-mail: kovacs.peter2@nje.hu

Abstract. Carbon fiber reinforced polymer composites are important and widely used material in today's manufacture and automotive. It produces a high demand to research the material, therefore industrial sectors can improve. Developing industry and strict requirements leads us to use this type of material in different applications where we need to use bonding technologies. The exploration of bonding technologies is necessary, and adhesive bonds offer a promising alternative. This paper examines the adhesive bonding of carbon fiber composites using a high strength two-component epoxy.

1. Introduction

Carbon fiber-based polymer composites stand out among various polymer matrix composites, significantly impacting different applications. This is due to the exceptional capabilities of carbon fibers: immense strength, high specific modulus, remarkable temperature resistance, superior electrical conductivity. Moreover, carbon fiber possesses unique characteristics like a remarkable strength-to-weight ratio, minimal environmental impact, non-corrosiveness, and exceptional durability [1]. Carbon fiber are widely used in different application areas like aerospace, automotive industry [2] Figure 1. shows the amount of research publications of carbon fiber during the last 20 years [3].

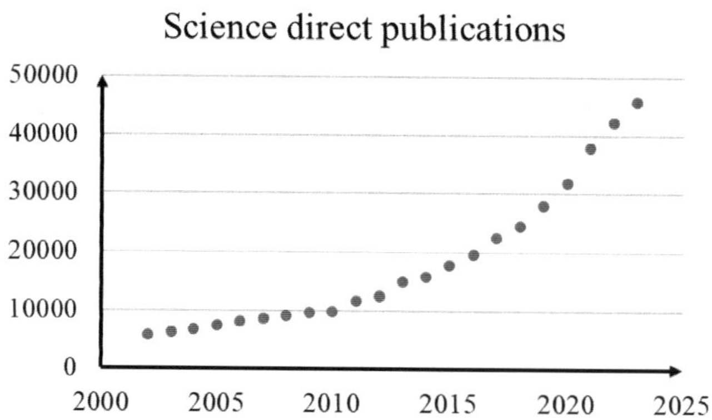

Figure 1.: Relevance of the research topic [3]

Bonding technologies play an important role in industry. Adhesive bonding can provide various advantages if you consider it in different applications. Therefore, it produces a need to develop adhesive

bonding technologies. Following this, I show the relevance of it, in different studies, that has addressed this issue.
J. Gebauer et al. offers a comprehensive analysis of a novel method for bonding carbon fiber-reinforced plastic (CFRP) with metal, particularly focusing on copper-coated FRP and steel. The authors propose an innovative process chain that incorporates laser pre-treatment of FRP before metallization through thermal spraying, followed by a low-temperature soldering process. This approach is designed to establish a reliable multi-material bond without harming the structural integrity of either the FRP or the metal part. The laser pre-treatment creates a rough surface, enabling deep interlocking of the thermally sprayed copper coating with the CFRP substrate, effectively doubling the adhesion strength of copper-coated CFRP to 18 MPa compared to grit-blasting. The performance of this hybrid material is impressive in various tests. It achieved a flexural stress of 1128.0 MPa in a 4-point-bending test without any coating impairment and reached up to 15.5 MPa in tensile shear tests. In terms of durability, the material withstood 5 million load cycles in fatigue testing at a maximum stress of 15 MPa and a mean stress of 8.6 MPa, showing no measurable structural damage. Additionally, the joining zone remained unaffected after exposure to saltwater [4].
M Berczeli et al. research focus is heavily placed on diminishing the mass of vehicle structures and their components, a move aimed at reducing harmful emissions and fuel consumption [5] They are developing a range of innovative techniques, notably the employment of diverse hybrid material combinations. These are selectively used based on the structural zones and the specific strength requirements of different parts. This methodology enables the integration of higher-strength, costlier metals solely in areas where mechanically essential, while in other regions, lighter metals and composites are deemed sufficient. A notable challenge in this domain is the bonding of elements with varied properties [6,7]. Here, adhesive technology is gaining attention as a more efficient alternative to traditional joining methods like riveting, welding, and soldering. Their findings are quite revealing: for instance, adhesions with alcohol-cleaned aluminium sheets demonstrated a 34% increase in strength (4662 N) compared to those with uncleaned sheets (3485 N). The strongest bonds were observed with sandblasted aluminium sheets, showing a remarkable strength of 5044 N, which notably exceeded the strength of bonds with polished (4985 N), etched (4850 N), or untreated (4662 N) sheets [8]. Additionally, their research highlighted the influence of surface treatments on adhesive performance. For example, in the case of polished aluminium surfaces, the strength of the bond was such that during tensile tests, a small piece was torn from the fiberglass fabric. In contrast, with sandblasted samples, the adhesive adhered perfectly to the aluminium, and during tensile tests, the entire layer of fiberglass beneath the bonded surface was torn from the composite [9].

The study investigated the effects of different surface treatments on the adhesive bonding strength of carbon fiber reinforced composites. They examined treatments like solvent cleaning, sanding, acid and alkaline etching, and peel ply removal. The study found that these treatments significantly influenced the adhesive bonding strength, as indicated by variations in interlaminar fracture toughness (GIC) values and delamination lengths. Notably, acid etching showed the highest GIC value (1.8 kJ/m^2), indicating strong bonding. Conversely, base etching resulted in the lowest GIC values (0.28 kJ/m^2), suggesting weaker bonding. The findings suggest that surface treatment is crucial for optimizing adhesive bonding in composite materials [10].
Adhesion behaviour of the reinforcement and the polymer influences the properties of the base material. It also affects the ability to absorb the load [11].

2. Equipment, methods, materials

The carbon fiber reinforced polymer composites were produced from 4 layers of carbon fibers embedded in epoxy resin. The composite plate has a thickness of 2 mm. From the composite sheet the specimens were cut with a plate shear machine. The specimens were 75 mm long, and 25 mm wide, to be able to apply the adhesive in the area of 25mm×25mm.
LOCTITE® EA 9466 adhesive was used to produce the specimens for the examination. This adhesive is a two component high strength epoxy adhesive with high shear strength.

The laser beam surface treatment was created using a Coherent Dilas Compact Evolution 300/200-980 LC diode laser beam source. A previously developed parameter setting was used to perform the surface treatment.
Grinding was created using a flap disc grinder. The grinding was applied transversely on the surface of the specimens.
For the sandblasting Torin Big Red DJ-SBC990 industrial sandblasting cabinet was used. The machine operated with quartz particles of 300μm. The sandblasting was done at an angle of 30-60 degrees.
The tensile test was made with INSTRON 5900R 4482 universal tensile tester machine.

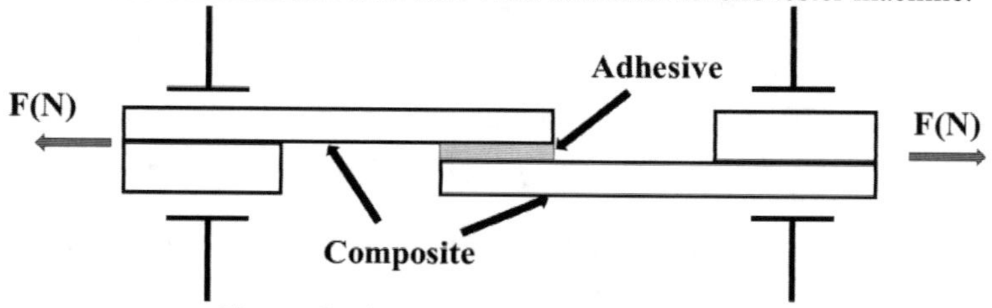

Figure 2.: Arrangement of the tensile test

To develop the bonds, we applied different surface treatment on the specimens. The examined sample types were control, grinded, sandblasted, laser treated, grinded and laser treated, Sandblasted and laser treated. Five specimens were used for each examined surface treatment.

3. Experiment

The Figure shows average force at the adhesive bonds break.

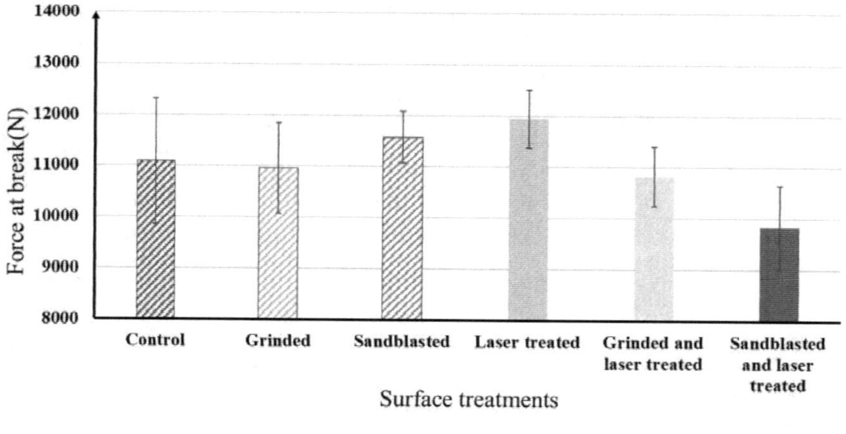

Figure 3.: Braking forces of the surface treatments

The following relationships between different surface treatment processes and the shear strength and dispersion of epoxy bonded joints of carbon fibre composites could be established, taking into account the material science aspects:

- Control: the control group is the reference point, where the mean shear force is 11078 N, and the standard deviation is 2457 N. The high standard deviation suggests that the consistency between the bonds may be variable, which may be a consequence of the lack of surface treatment.
- Grinded: the grinded specimens have a slightly lower average shear force (10959 N) but a significantly lower multiplication (1773 N) than the control group. Sanding produces a smoother surface, which can improve the bond between the adhesive and the composite, but the strength and method of sanding can significantly affect the bond strength.

- Sandblasting: Sandblasted samples show the highest shear force (11572 N) and a relatively low dispersion value (1014 N). Sandblasting results in a higher surface roughness, which can increase mechanical locking and thus bond strength, and can improve adhesion between the adhesive and the composite.
- Laser beam treated samples have a higher shear force (11936 N) than the control but lower than the sandblasted sample, but a medium dispersion (1136 N). The laser beam can modify the surface topography and chemical composition, which can affect the bond strength.
- Grinded and laser beam treated samples have a lower shear force (10825 N) compared to the control and a similarly low variance (1148 N). The combined treatment can produce a variety of surface characteristics, which can have complex effects on adhesive bonding.
- Sandblasted and laser beam treated samples have the lowest shear force (9837.8 N), but the scatter is the highest here (1635 N). This may suggest that the combined surface treatment methods do not give consistent improvements in bond strength and their interaction may be complex, increasing the variability of the bonding results.

Overall, the sandblasted samples showed the best performance in terms of shear strength, which is consistent with the material science principle that higher surface roughness can improve mechanical locking and adhesion. The lower scatter of the grinded and laser beam treated samples indicates that these surface treatment methods produce more predictable bonding results, although shear force may be reduced. Understanding the effects of different surface treatment methods is key to optimising bonded joints in carbon fibre composites.

The Figure shows the average force when the samples broke.

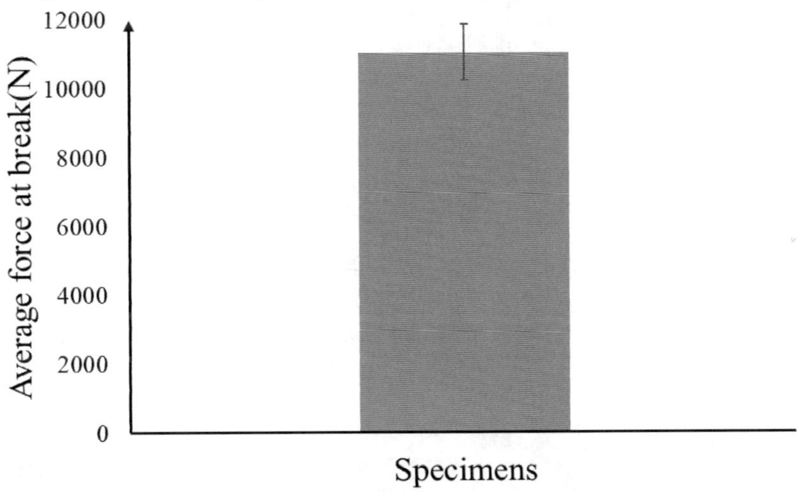

Figure 4.: Average braking forces of all specimens

When the laminating and internal failure of carbon fibre reinforced composites (CFRP) is the determining factor in shear tests, the material science context is significantly different from that which would be experienced when measuring bond strengths alone. Here, the following materials science considerations need to be taken into account:

Interlaminar failure can be seen laminating of the composites means that the tested bonds failed not because of adhesion failure, but because of weakness of the bonding forces between the inner layers of the CFRP material. This indicates that the bond between the carbon fibres and the resin is weaker than the bond between the adhesive and the surface of the composite.

Critical load of the CFRP material is indicating that the internal strength of the composite material is the determining factor, rather than the quality of the surface finish or adhesive.

There are other aspects which need to be investigated:

- Average Shear Strength: Average shear strength values indicate the destructive strength of the CFRP material.
- Standard Deviation: The standard deviation values indicate the consistency of the internal strength of the CFRP between different samples. A low standard deviation indicates a consistent internal quality of the material, while a high standard deviation may indicate variations in quality.
- Inherent Quality: The frequency of internal failure of CFRP indicates that the internal quality of the composite material should be improved, perhaps through carbon fibre arrangement, resin composition, or manufacturing processes.
- Surface Treatment Effect: The effect of surface treatments is negligible in this case, as internal degradation dominates. However, surface treatments can affect laminating indirectly, for example by changing the internal stresses caused by heat treatment.
- Bonding Technique: Since the bond strength depends on the internal strength of the CFRP, less emphasis should be put on optimising the bonding technique and more on the material quality.
- Material Investigations: Detailed investigations of the internal structure and manufacturing process of the CFRP material are needed to identify and correct weaknesses.
- Manufacturing Process Improvements: Fine-tuning of manufacturing processes, such as resin cure times and temperatures, can improve bond strengths between laminates.
- Design Optimal: Design of composites under load should consider internal failure points to ensure that structural elements do not fail due to internal weaknesses in the material.

The results of the shear tests show that the internal failure of CFRP material should be focused on in future developments. Optimisation of surface treatment and bonding methods may be important, but improving the internal quality of the composite material is necessary to avoid these types of failures.

The following figures show the surfaces of the specimens after the tensile test.

Figure 5.: Surface of the base material

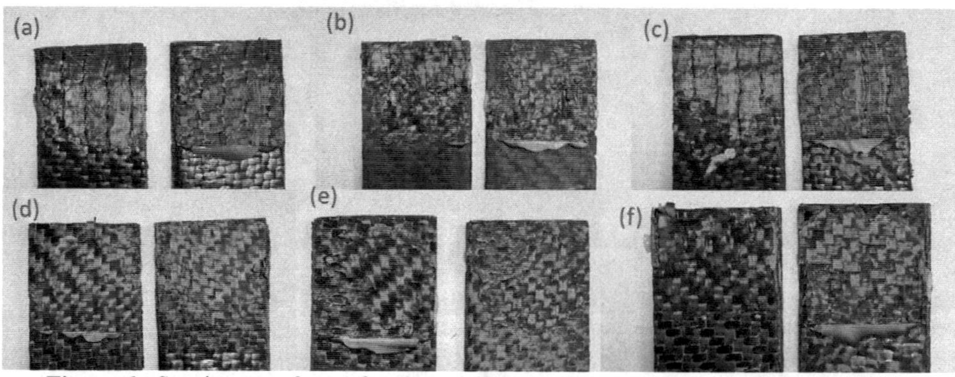

Figure 6.: Specimen surfaces after the tensile tests. Marked surfaces were (a)control (b)sandblasted (c) grinded (d)laser treated (e) sand blasted and laser treated (f) grinded and laser treated.

4. Conclusions

The article investigates adhesive bonding techniques for carbon fiber reinforced polymer (CFRP) composites, which are increasingly used in industries such as automotive and aerospace due to their high strength, lightweight properties, and durability. The study focuses on enhancing the adhesive bond strength using a high-strength two-component epoxy and evaluates the effects of various surface treatments, including grinding, sandblasting, and laser treatment. The findings indicate that sandblasting significantly improves the bond strength by increasing surface roughness, thereby enhancing mechanical interlocking and adhesion. Conversely, combined treatments like sandblasting and laser did not consistently improve bond strength, suggesting complex interactions between different surface modifications. The research underscores the importance of optimizing surface treatment processes to maximize adhesive bonding performance in CFRP composites.

Acknowledgement

This work was supported by the 2020-1.1.2-PIACI-KFI-2021-00260 Grant. The authors are grateful to the scholars cited in this paper, and their research results have given a lot of inspiration to this work. Special thanks to the staff of tour department for their support

5. References

[1] Y. Wang, A. Li, S. Zhang, B. Guo, D. Niu:A review on new methods of recycling waste carbon fiber and its application in construction and industry, Constr. Build. Mater., 367 (2023), Article 130301, 10.1016/j.conbuildmat.2023.130301

[2] R. Gogoi, A.K. Maurya, G. Manik: A review on recent development in carbon fiber reinforced polyolefin composites, Compos. Part C Open Access, 8 (2022), Article 100279, 10.1016/j.jcomc.2022.100279

[3] https://www.sciencedirect.com/search?qs=carbon%20fibre
Webpage visited: 2023.12.05.

[4] Gebauer, J., Drehmann, R., Eisner, K., Kupke, D., & Lasagni, A. F. (2022). Development of Multi-Material-Joint Made of Copper-Coated FRP and Steel by Innovative Technology Fusion. Technologies for Lightweight Structures (TLS), 5(1), 104–114. https://doi.org/10.21935/TLS.V5I1.169

[5] Berczeli M. Wettability Changing of FINEMET Substrates Using High-Energy Femtosecond Laser Impulses. Acta Phys Pol A. 2020;137(5):864-867 PG - 4. doi:10.12693/APhysPolA.137.864

[6] Berczeli M, Weltsch Z. Enhanced wetting and adhesive properties by atmospheric pressure plasma surface treatment methods and investigation processes on the influencing parameters on HIPS polymer. Polymers (Basel). 2021;13(6):901. doi:10.3390/polym13060901

[7] Berczeli M, Kürtösi R, Tímári A, Líska J. Optimize the joining technology of CFRP and aluminium sheets using hybrid and high strength adhesive. IOP Conf Ser Mater Sci Eng. 2018;448. doi:10.1088/1757-899X/448/1/012066

[8] Berczeli M, Weltsch Z. Mechanical properties of different strength steels MIG brazed joints. In: 34th International Colloquium on Advanced Manufacturing and Repairing Technologies in Vehicle Industry.; 2017:29-32.

[9] Berczeli M, Buza G. Relationship between the keyhole laser welding and the plasma. IOP Conf Ser Mater Sci Eng. 2018;448. doi:10.1088/1757-899X/448/1/012051

[10] Martínez-Landeros, V. H., Vargas-Islas, S. Y., Cruz-González, C. E., Barrera, S., Mourtazov, K., & Ramírez-Bon, R. (2019). Studies on the influence of surface treatment type, in the effectiveness of structural adhesive bonding, for carbon fiber reinforced composites. Journal of Manufacturing Processes, 39, 160–166. https://doi.org/10.1016/J.JMAPRO.2019.02.014

[11] Zhang, S., Hao, A., Nguyen, N., Oluwalowo, A., Liu, Z., Dessureault, Y., Park, J. G., & Liang, R. (2019). Carbon nanotube/carbon composite fiber with improved strength and electrical conductivity via interface engineering. Carbon, 144, 628–638. https://doi.org/10.1016/J.CARBON.2018.12.091

AUTHOR INDEX

Adonyi, Yoni 76
Backer, Jeroen De 76
Balazs, B Z 103
Bárány, T 58
Bereczki, P 18
Buza, G 120, 126
Capár, A 137
Czigány, T 10
Dessie, J E 42
Djimaoui, Nour El Imane 76
Dobránszky, János 68
Fábian, Er 50
Fábián, R E 120, 126
Fendrik, Á 111
Görbe, Á 58
Halász-Kutasi, I Z 58
Hareancz, F 120, 126
Horváth, R 50, 88, 131
Jacso, A 103
Juhász, G 120, 126
Kemény, A 1, 35
Kohári, A 58
Kohlhéb, R. 88
Kondas, B 94
Kovács, Árpád 76
Kovács, N K 10
Kovács, P I 144
Kuzsella, László 76
Lukács, Z 42
Májlinger, K 1
Maróti, J E 35
Marton, G Zs 111
Mertinger, V 94
Mertinger, Valéria 76
Oláh, F. 131
Orbulov, I N 1, 35
Pados, G 1
Rácz, I 10
Réger, M. 88, 131
Réger, M 50
Réti, T 50
Szebényi, G 111
Szederkényi, B 10
Szlancsik, A 28
Szovák, B 35
Tajti, F 137
Takacs, M 103
Tancsa, V 103
Tatár, L 18
Varga, D 28
Vero, Balázs 68
Vida, Á 120, 126

Institute of Physics
Dirac House, Temple Back
Bristol BS1 6BE UK

ISSN: 1757-8981
ISBN 979-8-3313-0660-1

9 798331 306601

CURRAN ASSOCIATES INC.
proceedings.com